Non-Destructive Testing

Non-Destructive Testing
and testability of materials and structures

Gilles Corneloup, Cécile Gueudré and Marie-Aude Ploix

EPFL PRESS

EPFL PRESS is an imprint owned by the Presses polytechniques et universitaires romandes, a Swiss academic publishing company whose main purpose is to publish the teaching and research works of the Ecole polytechnique fédérale de Lausanne (EPFL).

PPUR, EPFL – Rolex Learning Center, CP 119, CH-1015 Lausanne, info@epflpress.org, tél.: +41 21 693 21 30, fax: +41 21 693 40 27.

www.epflpress.org

© 2021, First edition, EPFL Press
ISBN 978-2-88915-440-1

Printed in Italy

FOREWORD

Non-destructive testing (NDT) encompasses many methods whose aim is to inspect an object without modifying it. NDT has to determine with suitable means whether a product conforms to pre-established specifications or requirements, and it must result in an unambiguous decision regarding the acceptance, scrap, or repair conclusion it has reached. Today's needs for NDT are tremendous and cover extremely diversified fields such as the aeronautic, petrochemical, automotive, nuclear, railway, and food industries. It has become impossible to overlook this activity over the past years.

What is generally compared is the evolution of the physical quantities between a part considered as safe, at the scale of the measurement, and another part with a defect. The presence of so many different non-destructive methods can be explained by the large diversity of the general characteristics of a part to be inspected: vital or non-vital mechanical parts, single part production or mass production, made by machining, welding, or bonding processes, etc.

The material can be metallic, ferromagnetic, or other, ranging from the scale of a millimeter to a meter, and may need to function in adverse environments (temperature, underwater, radiation, inaccessibility, etc.) Automated testing is sometimes necessary, and the important and permanent problem of supplying and saving (possibly over long periods of time) irrefutable evidence of the testing still remains.

Obviously, before any non-destructive testing is carried out, an essential matter is the choice of the most suitable method. The purpose of this book is to provide tools to meet this complex need.

Our wish in this work is to be exhaustive in the identification of the parameters to be taken into account to solve real NDT problems. These parameters are defined and only the necessary theoretical limits for choosing the optimal NDT method are presented. Further information will be made available for each NDT method in specific works to be published and will form part of a complete series called "METIS Lyon Tech."

Chapters 1 and 2 introduce the major parameters to take into account when choosing the best NDT or NDE (non-destructive evaluation) method. It is essential to have a good understanding of the part and the possible defect in order to carry out a relevant inspection. In this first section, the reader will become familiar with the logic that guides the choice of an optimal NDT method. The analysis of the conditions leading to the appearance of possible defects, in relation with the type of material or structure and the way they are obtained or damaged is comprehensively covered. Conclusions about material and structure testability are suggested.

The different NDT methods are then introduced in chapters 3 to 12. These chapters include the 10 inspection methods defined in the NF EN ISO 9712 standard, each one being symbolized by 2 letters, acronyms of the English designation (see Appendix A). Physics principles, technology, testing equipment, and implementation techniques are considered for each method and illustrated by different applications. Current advances in technology have led to a debate over the potential of these methods. Classic industrial techniques are addressed as well as methods corresponding to a specific need, specialized methods, methods with a limited use, and methods without a certification process.

Chapter 13 highlights the factors that can negatively affect NDT measurements, for example, a weakening in amplitudes, which may possibly distort the diagnosis. In most cases the basics on structural noise (and thus signal-to-noise ratio) are introduced through examples using ultrasounds, eddy currents, and so on. Environmental parameters such as temperature are also taken into account.

Chapter 14 focuses on solutions, in particular the progress that has been made in phenomena modelling and inversion methods. This chapter also outlines post-processing of data (signal, image) or even data fusion.

We will also see how taking into account NDT rules of good practice can result in better performance and cheaper testing costs: these are the RC-CND (*Recommandations de conception des structures basées sur les règles de CND* – Recommendations for structure design based on NDT rules).

The reader who wishes to go further into any particular points or the issues developed will find a bibliography listing the main documents to consult at the end of each chapter.

This work can be read at different levels: it is designed for readers who are discovering NDT, as well as for experienced professionals. For the latter category, level 2 professionals will find information necessary for their jobs, but this document will be of interest for all methods at levels 3 and "4" (a level which can be used for people in R&D) as a useful aid in making optimal choices for NDT.

The book will also be of great interest for students in bachelor, master, and doctorate studies in order to get a thorough overview of this line of research, understand its limits and potential for developing new NDT methods, and learn how to refine and use non-destructive experiments in their work.

This work is the fruit of collaboration among many specialists and experts in the field, including industry members, researchers, and university researchers. Some of them contributed with their writings (publications, texts from lectures, and often documents from initial or professional training courses). All of them were interviewed. There were several meetings where we exchanged points of view. All of this came together in the writing of this 14-chapter corpus, which takes into account the wide fields of knowledge required for each method together with a simple and clear editorial line.

For their personal involvement in this work, we would particularly like to recognize and thank the following people (specific collaborations are mentioned in the corresponding chapters), with a special mention for Jean-Michel FIEFFE, who passed away in 2016:

Francis CASADO	Cegelec NDT
Jean-François CHAIX	AMU/LMA
Michel DESCOMBES	INSAVALOR
Jean-Michel FIEFFE	SIRAC
Nathalie GODIN	INSA Lyon
Yves JAYET	INSA Lyon
Jean-Michel LETANG	INSA Lyon
Joseph MOYSAN	AMU/LMA

We also had many fruitful exchanges with other well-known personalities from the NDT field, may they be thanked, too, for the relevant talks we had:

Michel ALLOUARD	Snecma
Philippe BENOIST	M2m
Olivier CASSIER	Sofranel
Philippe DUBOIS	Extende
Jean-Claude GANDY	Sgs Qualitest
Philippe GUY	Insa Lyon
Pierre HUSARECK	Sofranel
Isabelle MAGNIN	Creatis
Thomas MONNIER	Insa Lyon
Marc PEYROT	Insavalor
Christian VENTURE	Sgs Qualitest

Many companies authorized us to reproduce the high quality photographs with which this work is illustrated. For this collaboration and for their support we are grateful to the following companies:

ACTEMIUM	AIRBUS INDUSTRIES
ANAÏS	AREVA
ATEM	BALTEAU
CEA	CETIM
CIE	CNDR La Réunion
DASSAULT AVIATION	COFREND
DUNOD	EDF R&D
EXTENDE	IFAT
CANADIAN ARMED FORCES	INDUSCOPE
GROUPE INSTITUT DE SOUDURE	KEM ONE
HOWDEN BCCOMPRESSORS	LAVOISIER
M2M	MISTRAS
NTB	OLYMPUS
OMIA	RTE
SAFRAN-SNECMA	SAUTER
SGS QUALITEST	SOFRANEL
SREM	TESTO
VIDISCO	YXLON
ZEISS	

As well as the following universities and schools: Aix-Marseille Université, INSA Lyon, Polytech Marseille (IUSTI Laboratory), IUT Aix-Marseille, IUT Nîmes, Mines Douai, Mines Paristech.

We would also like to thank COFREND, who authorized us to use their knowledge in the Non-Destructive Testing professional environment and helped us quantify in chapter 1 (NDT in context) the reality of a testing operator's job.

This writing relies on the expertise gathered in the studies carried out in the LCND laboratory (Laboratoire de Caractérisation Non Destructive – Non-destructive Characterization Laboratory) since its creation in 1989 in the premises of the Génie Mécanique et Productique (Mechanics and Products Engineering) department, Aix-Marseille IUT. For many years this laboratory was the only French university laboratory in an Institut Universitaire de Technologie and its main research line was "Non-Destructive Testing and characterizing of real materials and structures."

LCND permanent members and colleagues naturally made a very large contribution to this work: Joseph MOYSAN, Ivan LILLAMAND, Jean-François CHAIX, Vincent GARNIER, and Cédric PAYAN.

Thank you to all those who helped us work out the chapters, for valuable exchanges when determining the potential of NDT methods, or when we were looking for relevant photographs.

Photo of the LCND permanent members, from left to right, Vincent, Jean-François, Cédric, Marie-Aude, Gilles, Cécile, Ivan, Joseph. And Jean MAILHE, Sandrine RAKOTONARIVO, and Alban GEAY who joined us in 2013.

Since 2012, the LCND has been under the management of the LMA (Laboratoire de Mécanique et Acoustique, Mechanics and Acoustics Laboratory, UMR CNRS 7031), within the "Waves and Imaging" team, specializing in mechanical wave propagation in complex fluid and solid media. The scientific theme of the team is cross-sector and centered on modelling for waves generally speaking (digital, physical, mathematical simulation, high performance calculations) with applications in Non-Destructive Testing, medical ultrasounds, underwater and seismic acoustics, and seismology. In this context, the team has been developing research aimed at imaging, characterizing, and testing natural, biological, and manufactured media in a non-invasive way, with the use of waves.

This work is also the result of a long collaboration with INSA Lyon and INSAVALOR, the origin of which is to be found in the 1980s and 90s with one of the first doctorate theses in NDT (Gilles CORNELOUP) in the frame of a doctorate training focusing solely on NDT. Different research work was then carried out as a collaboration between Lyon-based laboratories (CREATIS, MATEIS, CNDRI, etc.) and LCND, and teaching was given during various INSAVALOR continuous training courses making it possible to offer some original training (immersion ultrasound, NDT signals, and image processing).

This work has been made in the context of the NDTVALOR platform, which brings together the skills of researchers and professionals specialized in NDT and NDE and with a particular focus on offering training, consulting, development, and valorization actions using non-destructive methods, and more generally spreading knowledge in this field. NDTVALOR is the result of the collaboration of INSAVALOR and PROTISVALOR, subsidiaries of INSA in the universities of Lyon and Aix-Marseille respectively.

Gilles CORNELOUP, Cécile GUEUDRÉ,
and Marie-Aude PLOIX

gilles.corneloup@univ-amu.fr
cecile.gueudre@univ-amu.fr
marie-aude.ploix@univ-amu.fr

CONTENTS

CHAPTER 1

NON-DESTRUCTIVE TESTING IN CONTEXT

Non-destructive testing (NDT) encompasses many methods whose aim is to inspect an object without modifying it. These methods can be classified according to the physical phenomenon they use: acoustics (ultrasounds, acoustic emission), radiation (visual, X-ray, -ray, tomography, neutron imaging, infrared thermography), material flux (penetrant testing, leak testing), and electromagnetic fields (magnetic testing, eddy currents).

Historically, trade in NDT was developed in association with nuclear activities (1950–1980), followed by aeronautics (1980–2000), when composite materials appeared. More recently (2000–2010), new demand had focused on concrete characterization (civil engineering structures and containment systems). These needs (Fig. 1.1) are still present as well as many other demands for industrial and R&D applications, some of them exceedingly complex.

Fig. 1.1 A few important needs for NDT

Each NDT method is a collection of different specialized techniques (for example, focused or laser-generated ultrasounds, etc.). What is generally compared is the evolution of physical quantities between a part considered as safe at the level of measurement and another part with a defect. The presence of so many different methods is justified by the fact that the general characteristics of the object to be tested can be very diverse.

These parts can be vital or non-vital mechanical ones, from single part production to mass production, made by machining, welding, bonding, etc. The materials can be metallic, ferromagnetic, range from the scale of a millimeter to a meter, and the environment can at times be adverse (temperature, underwater environment, radiation, inaccessibility, etc.). Automated testing is sometimes necessary, and the important and permanent problem of supplying and saving (possibly over long periods of time) irrefutable evidence of the testing still remains.

The various methods of NDT typically have two objectives that are different but often complementary: research into macroscopic defects at the scale of the measure (this is what we will call NDT throughout this document) and the overall characterization of materials or parts (in which case the term **NDE non-destructive evaluation** will be used).

In the first case, the quality of the part is guaranteed by evidence of the absence of defects, whether microstructural or not. The testing can be executed in several steps: after the indispensable detection of events likely to be defects, it can be necessary, after an analysis, to identify them (location, surface or volume shape, dimensions) in order to deduce their level of noxiousness.

In the second case, testing is focused on some overall characteristics of the material, such as elastic response, stress conditions, homogeneity, or geometric parameters such as the thickness of a part or of a superficial layer. These tests can be executed by following the relative evolution of a physical parameter, as in defect research, but most often an absolute measure is necessary.

Prior to testing, it is essential to **choose the optimal NDT method**, that is, the most suitable method for a given problem. When testing for defects, the choice must take into account the general characteristics of the "part-defect" couple as well as those of the environment.

Ultrasonic testing is often used because of some of its intrinsic advantages, such as an easier implementation, the fact that it is not necessary to have access to both faces of a part (Fig. 1.2), good adaptation to the natural orientation of most defects, etc. But there are a few major drawbacks too, such as the need to couple the transducer to the part or the very high sensitivity of ultrasonic propagation to the various degrees of material heterogeneity or anisotropy.

In a context where quality necessitates searching for and identifying tiny defects on ever more complex structural materials (austenitic steel, composite materials, concrete, etc.), these intrinsic limits can be the cause of sometimes insuperable difficulties

Fig. 1.2 Ultrasonic inspection of an airplane wing (SOFRANEL)

in testing. The expression *structural material* here refers to real materials, the testing response of which is not equivalent to that of a perfect material. This convention is typically used in the NDT field.

Defect detection can also be difficult or impossible because of a degrading signal-to-noise ratio that contains information.

In such conditions it is no longer possible to find the shape and dimensions of the defect and the quality of the part cannot be proven. This is why an oversizing of the structure must be introduced in the design stage. Unfortunately, this is not enough to eliminate the problem in the case of a progressive defect, the problem is only delayed. This general problem of detection mentioned in regard to ultrasonic testing is to be found in all NDT methods.

1.1 NDT AND QUALITY CONTROL

Non-destructive testing is one of the methods enabling an appreciation of the *quality* or of some characteristics of an object, without affecting its integrity. The terms related to quality are defined in standards, but it is simpler to say that quality leads to the creation of a state of trust between partners, which gives them the assurance that the products are reliable without having to repeat tests already carried out by others.

This assurance may be based on a manufacturing process made in accordance with procedures that are clearly defined and that guarantee the absence of defects. Such precautions can be sufficient, especially in serial production, but in many cases quality control is necessary when manufacturing is more delicate (welding, for example), and/or because the required safety level is such that the quality must be proven (aeronautics, nuclear industry), and/or the cost defective work is too important (all fields).

Quality inspection has to determine, with appropriate means, if a product complies with pre-established specifications or requirements, including having to decide on acceptance, rejection, or repair. Quality control is mainly achieved thanks to non-destructive testing.

1.2 NDT AND MAINTENANCE

The European standard EN-13306 defines maintenance as the "combination of all technical, administrative and managerial actions during the life cycle of an item intended to retain it in, or restore it to, a state in which it can perform the required function." Maintenance policies have thus been set up, including:
- Maintenance based on reliability, with the purpose of maintaining the intrinsic reliability defined on design
- Maintenance based on risks, aiming to ensure that these risks remain to a level judged acceptable in the case of the failure of equipment

In the first case, non-destructive testing can be carried out systematically on all the parts, but if the safety goal is achieved, the cost can sometimes become excessive. It is also possible to implement an inspection method that uses sampling, which can show a shift in manufacturing. These methods can be empirical and/or statistical, using SPC (statistical process control) methods or standards specific to this type of inspection.

In the second case, regarding new approaches, non-destructive testing is now considered to be an element contributing to *AIM (asset integrity management)*. These concepts, especially that of major technological risk essentially appeared after the 1976 Seveso disaster. It represented a three-fold break with the past: technical (an accident may have effects beyond the industrial premises), economic (an accident may cost much more than the cost of maintenance), and socio-political (local populations also have a say).

After other major accidents (Amoco Cadix 1978, La Mède 1992, AZF 2001, etc.) the *RBI (risk-based inspection)* approach has finally become widespread. Risk-based inspection was initiated in 1993 by a consortium of 21 oil and petrochemical companies that desired a methodology (the RBI) for applications in refining and petrochemicals.

API (American Petroleum Institute) standards define the steps of the RBI inspection method. Their sole field of application is maintaining the mechanical integrity of *pressurized equipment* such as containers, pipes, vapor generators, etc., and their goal is to minimize the risk of containment loss due to damage.

This approach bases the inspection (including non-destructive testing) on the risks. Equipment (pipes, pressurized accessories, etc.), is grouped by iso-degradation loops based on similarity in degradation modes as well as in kinetics. They are the subject of a risk study defined by the seriousness and probability of the events.

The purpose of the risk study is to define the consequences of a failure, for people and environment, as well as for the company. To define this seriousness, the study of the potential hazards is based on three criteria: consequences on humans, the environment, and on cost. For each criterion, three or four parameters are chosen: inflammability, toxicity, domino effect, etc. from which a *seriousness index* can be estimated.

In the probability study, which represents the likelihood of a failure occurring, each mode of degradation is integrated (corrosion, creeping, etc., with feedback from past experience taken into account). A probability index is calculated based on the estimates of the degradation kinetics, from remaining service life (failure-expected-after-x-inspections) and from potential testability (easy or difficult inspection, etc.).

From these two values a *risk matrix* (Fig. 1.3) is drafted for the loop of iso-degradation. The zones of the matrix will indicate whether a risk reduction (from the source) is necessary and will enable the definition of inspections to reduce these risks. The risk reduction will be carried out by establishing and implementing inspection plans that define the nature (the NDT method used), the location, the span, and the periodicity of the inspections.

Determining the most suitable NDT method for a given problem (the main goal of this book), including all the parameters about the part, the defect, the environment, and so on, is thus an important element of this approach.

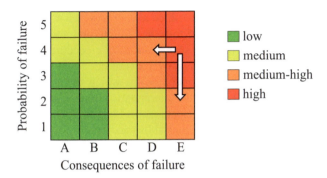

Fig. 1.3 Risk matrix and illustration of the 2 directions of risk reduction

The value of an RBI approach is to focus the means on the most "risky" equipment. It is now in use in the French state-run ***plan for the modernization of industrial installations PM2I*** (*plan de modernisation des installations industrielles*), for installations with significant quantities of hazardous products on the production site.

1.3 COFREND CERTIFICATION

Historically, the American requirement for quality assurance, particularly in the nuclear industry, is at the origin of the establishment of a personnel certification system enabling the industrial application of non-destructive testing in conformity with regulations, codes, standards, and specifications related to the inspected products. This certification improved the quality level of the inspections and made NDT a specific industrial activity.

This certification has to be made by authorized organizations. In France, the situation is quite clear, the ***COFREND*** (*Confédération Française pour les Essais Non Destructifs – French Confederation for Non-Destructive Testing*) was authorized in 1967 and delivers almost all the certifications. A few needs are covered by the American certification SNT from the ***ASNT*** (Association for Nondestructive Testing).

The certification covers competence in one or several of the following methods: acoustic emissions, eddy currents, leak testing (hydraulic pressure tests being excluded), magnetic testing, penetrant testing, radiography, ultrasounds, and indirect visual inspection (direct visual inspection with the naked eye and visual inspections executed within the application of another non-destructive method are excluded).

Both the COFREND and ASNT certification procedures thus offer three ***levels of qualification*** for operators, called ***levels 1, 2 or 3***:

- A level 1 certified operator is qualified to perform non-destructive tests based on written instructions and to record the results;

- A level 2 certified operator is qualified to choose which non-destructive testing technique is to be used for the test, then to interpret and evaluate the results according to standards, codes, specifications or procedures;
- A level 3 certified operator has proved his or her competence to execute and run any non-destructive test operation for which he or she is certified. The operator's competence covers the interpretation of results in terms of standards etc. and is combined with sufficient practical knowledge in materials to be able to choose the methods and help establish acceptance criteria when otherwise there is no available criterion.

These qualifications are obtained after an exam in each NDT technique, by *sector of activity* (energy-petrochemicals: CIFM; aeronautics: COSAC; metallic products: CCPM; railways: CFCM). They are valid for 5 years and must therefore be renewed. At the end of 2015 in France, there were 14,691 COFREND certified operators and 25,845 valid certificates (with some operators being certified for various methods). Table 1.4 outlines the principles of the French recommended levels for following a training course and the required experience necessary to apply for it.

Table 1.4 Principle of the 3 certification levels in France (see details in Appendix A)

	Recommended educational level	Required experience	Training	If success
Level 1	Vocational training certificate	1 year	2 weeks	Certificate valid for 5 years
Level 2	High-school diploma	6 months	2 weeks	Certificate valid for 5 years
Level 3	2 year's higher education	3 months	2 weeks	Certificate valid for 5 years

Time periods are only an indication, as the real certification conditions of attribution detailed in ISO standard 9712 represent a COFREND document of 39 pages. Appendix A "qualification and certification" in this book gives further information about this.

The distribution per level of certified people in 2015 shows disparity between the sectors of activity, the average being 15% for level 1 certification, 80% for level 2, and 5% for level 3.

COFREND certification is individual, though it must be approved by the employer. An ASNT certification, however, belongs to a company, which then delegates the task of conducting examinations for levels 1 and 2 certification to a level 3 operator. The choice comes from needs expressed by codes used in manufacturing, like CODAP, RCC-M, etc. for COFREND, and ASME, etc. for ANST.

For example, the AFCEN code RCC-M concerns mechanical equipment designed and manufactured for pressurized water reactors (PWR). It applies to equipment subjected to pressure in nuclear units and to a few components not subjected to pressure. In the same way, the ASME code covers industrial and residential boilers as well as components for nuclear reactors, transport tanks, and other kinds of pressurized equipment.

Apart from France, there are many organizations in other countries (almost one per country), but a strong presence of the ASNT is to be noted. Different guides for good practice in accordance with the prescriptions of ISO standard 9712 (specifying the requirements for the qualification and certification principles of the personnel in charge of industrial non-destructive tests) help harmonize the certification criteria.

1.4 THE NDT MARKET

A few reviews show the importance of NDT in the world, per industrial activity, per NDT method, per country, etc. Figure 1.5 presents the distribution in France, between the industrial inspection activities, R&D, materials production, and sale and training. KATALYSE, a research consultancy company, believes that the strong development observed between 2007 and 2012 will keep expanding because on one hand, the necessity for ensuring the continuation of ageing industrial structures has increased NDT needs and, on the other hand, the increasing complexity of industrial products and the growing demand for quality and risk management has favored the expansion of the NDT field of application.

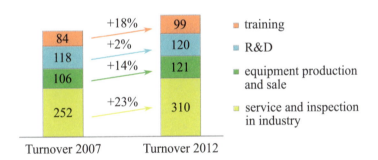

Fig. 1.5 NDT turnover evolution in France (in millions of euros), totaling €650 million in 2012 (KATALYSE estimates)

However, NDT equipment manufacturers believe these figures to be optimistic; according to them, NDT activity seems to have slowed down since 2012. Aeronautics is a rather stable market, as is the case for the automotive industry, petrochemicals, and railways, though the nuclear industry has not (yet) recovered its past importance. Electricité de France is a particularly important business – in some years it was the world industry leader in NDT, with almost 2% per delivered kilowatt/hour dedicated to NDT.

It is to be noted that a reason for the drop in NDT equipment sales is that current equipment is increasingly reliable, which can explain why technological developments like passing to phased array in ultrasonic detection (and other advances) are to be expected.

The distribution of 2015 COFREND certifications per activity sectors and techniques (Fig. 1.6), shows that four methods dominate the market, two for emerging defects exclusively, (penetrant testing – PT and magnetic testing – MT), and the other two for defects located inside the volume of the part (radiographic testing – RT and ultrasounds – UT).

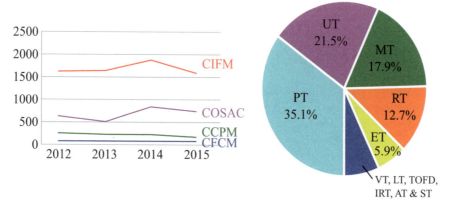

Fig. 1.6 Distribution of 2015 COFREND certifications per activity sector (energy-petrochemicals: CIFM; aeronautics: COSAC; metallic products: CCPM; railways: CFCM) and non-destructive methods

The figures in France conform to world trends as shown in the MRRSE review on the NDT equipment market (US$3.2 billion in 2014):
- Methods: ultrasounds, radiography, magnetic testing, visual
- Domains: energy production, aeronautics and defense industry, automotive industry
- Geography: USA, Europe, China, Japan, India

The service market is also important, and companies from this sector can be found in the top 20 of companies with the most valid certificates (Fig. 1.7).

In concrete terms, the cost of NDT services varies according to companies, used techniques, quantity, etc. but a few indicating costs can be given for 2016 (including traveling expenses and reports):
- Visual inspection: €350 to 450 per day
- Penetrant testing, magnetic testing: €400 to 500 per day, which represents testing for approximately 30 linear meters (LM) of welding, for example
- Ultrasounds, thickness measure: €400 to 500 for 300 to 400 measure points
- Ultrasounds, defect testing: €500 to 700 per day representing approx. 30 LM
- Ultrasounds, TOFD method, or with phased array method: €650 to 850 per day (it is possible to reach 40 LM when using PA)
- Radiographic testing (a more expensive method because there are often 3 people involved – 2 night-shift operators and 1 day-laboratory operator to develop

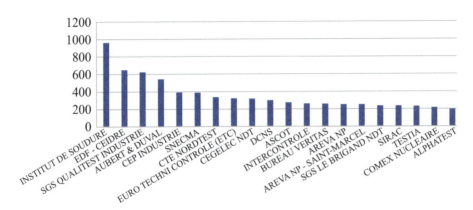

Fig. 1.7 France's top 20 companies with valid certificates (COFREND)

the films): €1,600 to 2,100 per day, representing 30 films (on site) to 40 films (in workshop), approx. 10 LM for 30–40 cm films

- ACFM, which is a complement to magnetic testing when paint thickness is significant: €750 to 900 per day, with over 50 LM
- Eddy current: €750 to 900 per day, with the possibility of testing, for example, about 300 tubes 4–5 LM
- Infrared thermography: €750 to 950 per day
- Acoustic emission: the cost of the service depends on the number of channels to be installed. For simple inspections (1–2 channels), the cost is about €1,500 per day, and may amount up to several thousand euros per day for systems with 50–100 or 200 channels (with many technicians working together)
- Leak testing: approx. €1,000 per day for standard test using helium
- Neutron imaging: this highly specialized technique (see chapter 6), is billed at €400 to 1,000 per film

These costs of industrial services do not concern complex inspections required by certain specific materials with high mechanical characteristics. A few other demands are not easy to satisfy, such as inspection, analysis, and diagnosis automation, or the control and processing of very large NDT data volumes (mass processing).

These special needs (complex materials, automation, etc.) are instead handled through research activity (and development), and French research, which accounts for a large proportion of all research, is recognized worldwide. In France, in each main activity sector (mainly nuclear and aeronautics), industrial laboratories coexist with university laboratories. In general, university laboratories have orientated their research around a single NDT method (quite often ultrasound), and this is the reason why the LCND laboratory approach, put forward as soon as it was created to deal with the general issue of testing without prior assumptions about the tool, is an original one and corresponds to specific needs.

1.5   REFERENCES

AFNOR, *Essais non destructifs, tome 1 certification du personnel et vocabulaire*, collection of Afnor standards, 2000.

AFNOR, *Essais non destructifs, tome 2 normes d'application générale*, collection of Afnor standards, Vol. 1, Vol. 2, 2000.

AFNOR, *Essais non destructifs, tome 3 normes d'application spécifique*, collection of Afnor standards, Vol. 1, Vol. 2, 2000.

API 571, Damage Mechanism Affecting Fixed Equipment in Refining Industry, American Petroleum Institute, December 2003.

CAPO A., *Introduction au métier d'inspecteur technique*, ISIM professional bachelor degree lecture document, an ARKEMA- IUT Aix partnership, 2014.

COFREND, Conditions d'Attribution des Certifications COFREND selon Norme ISO 9712, COFREND document, http://www.cofrend.com.

IDDIR O., Pondération des fréquences de fuite dans le cadre des analyses de risques industriels, *Techniques de l'Ingénieur*, Ref. SE5080, January 10, 2011.

ISO 9712, Essais Non Destructifs, Qualification et certification du personnel, ISO standards, 2012.

KATALYSE, La filière «Contrôle non destructif»: synthèse de l'étude des besoins de professionnalisation, KATALYSE document, http://www.katalyse.com, March 2014.

KOZLOWSKI A., Code de bonne pratique pour l'approbation des personnels d'essais non destructifs par des organismes tierce parties reconnus dans le cadre de la directive "Equipements sous pression," AFIAP meeting, October 22, 2008.

LE GOFF X., Bilan de la certification COFREND, COFREND, http://www.cofrend.com, 2015.

MRRSE Market Research Reports Search Engine, Non-Destructive Testing Equipment Market - Global Industry Analysis, Size, Share, Growth, Trends and Forecast 2015–2021, MRRSE report, http://www.mrrse.com/nondestructive-testing-equipment, July 14, 2015.

VENTURE C., L'inspection et l'état, le module RBI, l'inspection basée sur le risque, ISIM professional bachelor degree lecture document, an SGS QUALITEST- IUT Aix partnership, 2015.

ZWINGELSTEIN G., Méthodes de maintenance basées sur la fiabilité et sur les risques, *Techniques de l'Ingénieur*, Ref. SE1650, July 10, 2015.

CHAPTER 2

CHOOSING AN OPTIMAL NDT METHOD

In this chapter, we will first introduce a number of parameters that are to be taken into account when choosing the best NDT method for a given problem. This approach is not exhaustive, only the essential points that can demonstrate the logic guiding the selection of an optimal NDT method are considered.

Then, a few representative examples will show how the knowledge of the material and/or the structure can (and must) help to overcome certain complex difficulties in non-destructive testing. The testability of materials and structures is thus defined.

2.1 PARAMETERS WITH AN INFLUENCE ON THE CHOICE OF A NON-DESTRUCTIVE METHOD

The main goal of NDT is to search for local defects. In this section, defect characteristics are presented; they are extremely variable and mainly depend on the nature of the tested part, its mode of manufacturing, and how it operates. The same part can be tested several times during its existence. The influent parameters such as the presumed nature, location, and orientation of the defect, completely govern the analysis capacities of the NDT methods.

These multiple variables justify the necessity for having many NDT techniques at hand to meet the imperatives related to the numerous "defect-part" couples that are tested, whether in terms of defect detection, location, identification, or dimensioning.

And finally, we will deal with the problem of the global non-destructive evaluation of a part or a material.

2.1.1 General characteristics of the tested part

Tests are mostly carried out on steel structures and composite materials. The first case essentially concerns heavy industries (welding problems, etc.) and the nuclear industry (stainless steels, particularly austenitic and austenoferritic), and the second case is to be found in the aeronautic industry (monolithic composites, sandwich and bonded structures, etc.). The concrete used in bridges, dams, and containment structures also represents important problems for NDT.

The choice of an NDT method must take into account the various characteristics of the part: elastic (for acoustic methods), electrical (for eddy currents), attenuation laws (for radiography), etc. The anisotropy of a material directly influences the performance of the ultrasonic testing: depending on the propagation direction, the attenuation laws and deviation angles are modified. These influent parameters can evolve according to treatment (thermic, mechanical), temperature, span of damage, etc.

The size of the part can be of the order of a millimeter (thickness of vapor heat exchanger tubes) or a meter (diameter of rolling mills). The part can be inaccessible or it may be preferable to leave it in position. The environment can be hostile, as is the case in underwater testing, in zones exposed to atomic radiation, and high temperatures.

2.1.2 General characteristics of defects created when the part is manufactured

When a part is cast, it is a favorable moment for defects to appear. The passage from solid to liquid state with mechanisms such as retraction, dilution, or insertion can cause the apparition of *shrinkage cavities*, *blowholes*, *inclusions*, and more rarely, cracks (Fig. 2.1). Precautions must be taken on the design of the shapes, which must be regular, and on the execution of the casting operations (vents, deadheads, etc.).

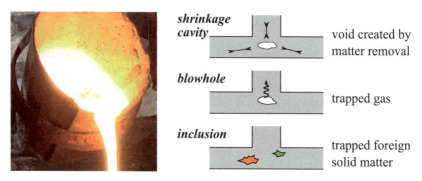

Fig. 2.1 Main defects in cast metal

A finer analysis is often necessary to identify the appropriate NDT method. In the case of a weld, for example, the fact that some cracks are caused by the combination of heavy clamping, a large fusion pool, and the presence of sulfur or phosphorus must be taken into account. These cracks appear around 1000°C (*hot cracking*), are located in the molten zone area, and are not generally ramified (Figs 2.2 and 3.10).

These cracks are not related to *cold cracking* (which appears around 300°C) located in the heat-affected zones and resulting from the coexistence of martensitic structure (quenching), stress concentrations, and the possible presence of hydrogen.

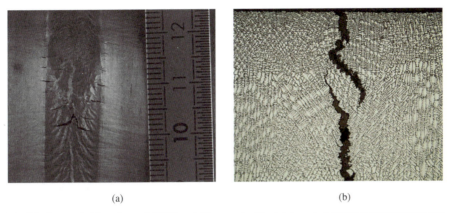

(a)   (b)

Fig. 2.2 Hot crack (a) on Inconel (EDF) and (b) on austenitic stainless steel (GROUPE INSTITUT DE SOUDURE)

Understanding which precautions to take at the moment of welding (basic electrodes, with a low level of hydrogen, and/or post-heating which lets hydrogen out) is vital for evaluating the risk and the type of cracking.

Sometimes, *delayed cracking* may happen 1 or 2 days after welding, resulting from hydrogen supersaturation in the dislocations of the crystalline network. Other types of welding defects can also happen: lamellar tearing, lack of fusion, lack of penetration, shape defects.

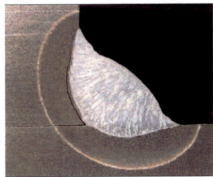

Fig. 2.3 Cold cracks under welding bead (GROUPE INSTITUT DE SOUDURE)

In the case of composite materials, the material production phase does not take place before that of the part but coincides with it. For monolithic composites, the presence of porosity, *delaminations*, and inclusions is feared; the control over some characteristics is later sought, such as *fiber rate heterogeneity*, their difference of orientation, and the waviness of the plies. *Fiber-matrix decohesion* is another defect for which detection is currently being developed.

In sandwich and bonded structures, the defects that are looked for are possible debondings, wrong insert positions, or the collapse of "honeycomb" structures. Testing means increased efficiency; the focus is no longer only on the non-conforming layers of adhesive, it is now on the problem of imperfect *adhesion*, even in the presence of adhesive.

The case of concrete is particular because this material rapidly evolves over time. It is sometimes difficult to distinguish between the defects that occurred during the production stage (structure heterogeneity, porosity, wrong filling of the pre-stressing ducts, etc.) from the defects that occurred later in time, but that were nevertheless caused by the production process. The *alkali-silica reaction* phenomenon, for example, is a reaction between silica aggregates and an alkaline solution that starts as soon as the concrete begins to set, but in some cases the mechanical properties of the material can take twenty years to degrade.

2.1.3 General characteristics of defects created during the service life of a part

Here we consider crack-type defects corresponding to ductile ruptures due to overload, fragile ruptures under generalized constraints below the elasticity limit, or fatigue ruptures due to a large number of cycles, low-cycle fatigue, or high temperature creeping. These rupture processes can be coupled with chemical processes (phenomena of corrosion under stress, for example).

Fatigue is the damage caused by stress with a cyclic character exceeding a threshold that leads to a progressive cracking of the parts. A crack is generally initiated on the surface, at the location of a notch (whether geometrical or metallurgical), creating a local stress concentration. Sub-surface or internal crack initiation can be observed in the presence of (non-detected!) production defects.

Creeping is a slow deformation that is continuous in time and dependent not only on constraint but also on time and temperature. Damage is caused by multiplication and widening of internal cavities (Fig. 2.4), appearing at grain boundaries. After the accumulation and coalescence of the cavities, the final rupture occurs. As the presence of grain boundaries is detrimental, materials with orientated solidification

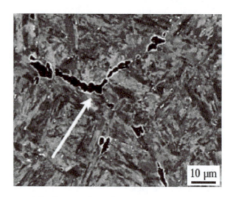

Fig. 2.4 Creep damage with micrometric cavities (arrow) (V. Gaffard, MINES PARISTECH)

or monocrystalline materials have been developed, minimizing the number of grain boundaries, though raising other types of testing problems.

Corrosion may cause general degradation, and inspection can reveal a reduction in thickness or local damage (pitting corrosion, crevice corrosion). Intergranular corrosion concerns austenitic and ferritic stainless steels; they may undergo selective corrosion at grain boundaries leading to a total disaggregation of metal without measurable mass loss. Stress corrosion, which occurs in the presence of mechanical stress of diverse origins (external or residual, especially in the case of welding) may cause severe trans-granular cracking.

2.1.4   Defect noxiousness

Chapter 1 introduced RBI-type probabilistic methods of risk evaluation. The analysis of the various seriousness factors is here extended by the specific study of *defect noxiousness*, as understood in fracture mechanics, which is mainly linked to the shape and size of a defect.

Volume defects (blowholes, inclusions, or even triple-point creeping) do not intrinsically have a predictable evolution, however, they may become the origin of a cracking phenomenon. This is why NDT is used to focus on their total number, their distribution in a part, and the dimension of the largest one. In the industrial context of part quality measure, the presence of volume defects is thus tolerated if their quantity and/or maximum dimensions remain under a limit value.

In principle, *non-volume defects* are not accepted because they propagate and will in time cause the rupture of the part. According to this binary logic (defect/no defect), the part must be rejected or repaired. But a refined analysis shows that for evolving defects (for example, fatigue cracks), it is generally possible to distinguish the initiation phase from the propagation phase and sudden or unstable propagation from stable propagation.

At the industrial level, a sudden rupture is evaluated by NDT only in a context of inspection-repair, when evidence that it happened on the elements of a complete structure needs to be provided. But in the case of rupture by successive cracking, the notion of defect critical size has been demonstrated: a size beyond which, for a given stress, an unstable rupture is initiated.

Different works have shown that the binary criteria used in quality inspections are often too severe (finding a crack leads to repair or rejection) in terms of *tolerable maximum defect* size (Fig. 2.5). As a consequence, from real noxiousness calculations

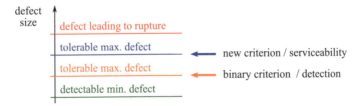

Fig. 2.5 Tolerable defect sizes from a serviceability viewpoint

taking into account critical sizes, it is possible to introduce the concept of product *serviceability*, where service behavior is correct throughout the specified service life.

This makes it possible to raise the size threshold of the evolving defect, but periodic non-destructive testing will be necessary to guarantee the quality of the part, proving either that the defect has not appeared or that it has not reach a critical size.

Defect noxiousness also depends on its *location* and its *orientation*. The estimated geometric situation for the indication of a defect (in ultrasounds, a defect is first called a "reflector") gives an idea about its nature: for example, slag inclusion cannot be found outside the molten zone of a weld.

Prior knowledge of the orientation of a defect helps in the selection of the most suitable NDT method. Ultrasounds and radiographic testing, for example, are based on the respective principles of reflection and attenuation, which enable the detection of plane defects having "complementary" orientation, either perpendicular to the ultrasound beam or parallel to the photon beam.

The most precise determination of the size of a crack by NDT then becomes part of the predictability of the service life of a structure. We can then understand how indebted fracture mechanics (essentially focused on the study of macroscopic cracks and the behavior of cracked parts) is to progress made in the development of NDT methods.

2.1.5 Representation of information, diagnosis, and evidence

Information obtained after a non-destructive test is generally represented in the form of a 1D signal (ultrasounds, eddy currents, etc.) or a 2D image (radiography, penetrant testing photographs, etc.). Sometimes, a 3D data cube is generated (tomography, ultrasounds, etc.). When a choice is possible, it can be difficult to recommend a single mode of representation, and the display of multiple images is often used.

Figure 2.6 shows a set of BScan, DScan, and CScan images (green backgrounds) reconstructed from AScan ultrasonic signals (on the right, in the middle). The ultrasonic amplitude of the AScan is in ordinates (yellow axis), in the time function (blue axis). These images position the acquired information onto a specific reference system. These modes of representation will be explained in chapter 7, but it is already clear that ultrasound images, although useful for presenting and archiving the inspection, are sometimes too complex to allow a diagnosis to be easily reached. This is why the original simple AScan signals are often preferred for measurements and threshold comparisons.

The notion of archived evidence of a test is important in the case of the natural or accidental evolution taking place on the tested object. What was its original condition? In this context, often with legal issues, the choice of an NDT method enabling archiving over a long period of time (several dozen years) that leaves no question of fraudulent modifications can be an important criterion.

Radiographic testing on analog film, for example, today guarantees a legally indisputable "zero point" as opposed to digital information. The current durability of analog film is still acceptable even if its industrial production has greatly decreased, due to the shift to digital systems in photography.

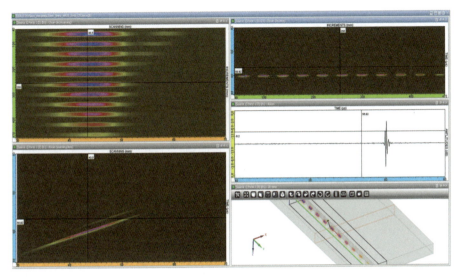

Fig. 2.6 Ultrasound images of a part with artificial defects, with axes marked by colors (EXTENDE)

2.1.6 Automation, rate

The question of whether to automate an NDT method is an important choice criterion that is naturally found in the case of serial testing and in hostile environment testing (temperature, radiation, off-shore), when a robot has to take the place of the human operator.

Ultrasound testing, which in many applications performs so well, is disadvantaged in automated operation by the necessity of coupling the transducer to the part (by liquid, gel, etc.). Testing without coupling (air, laser, EMAT, etc.) is possible but currently less efficient.

When possible, electrical methods such as eddy currents are of the most interest. They need no coupling, are ultra-fast, and the acquisition system drift along time is near zero.

2.1.7 Environmental impact

Methods using fluids and other fine powders (penetrant testing, magnetic testing) can be used once the problem of recycling the products has been settled. The use of developers and liquid supports can be hazardous for the environment and health, which is why their use, conditioning, storage, and recycling must be carefully detailed. All this can be more or less restrictive and may become a criterion in the choice of an NDT method.

Similarly, considering the high activity of the radioactive sources used, industrial radiographic testing represents a risk to workers that must be prevented and limited. Any exposure to ionizing rays has to be justified by the advantages compared to the risks. Resorting to technologies with lesser risks has to be favored: this is the principle of justification for radiographic testing.

2.2   EXAMPLES OF NON-DESTRUCTIVE EVALUATION (NDE)

NDE is not about finding a possible macroscopic defect (with location, shape, and dimensions) as evidence for the quality of a mechanical unit, but the global evaluation of one of its mechanical, physical, or geometrical characteristics.

There are few non-destructive methods enabling this kind of evaluation, so the issue of selecting which one to use becomes much simpler.

The characteristics that can be measured in a non-destructive way are nonetheless very diverse.

- Material damage due to the presence of microscopic defects (micro-cracks, micro-cavities). In this case, a problem of scale prevents each elementary defect from being identified (Fig. 2.7), but it is possible to try to determine if this damage can be found through a modification in the laws of attenuation (ultrasounds, photons), in electrical conductivity, etc.
- The mechanical characteristics of a material (elasticity constants, anisotropy, hardness) or a system (stress in a bolt, adhesion of an adhesive). Methods based on elastic wave propagation are mainly used for this purpose. The hardness of steel, however, can be tested by eddy currents, because the electrical conductivity of a material is expressed as a function of the size of the grains of structure, which is itself linked to hardness.
- Product homogeneity (distribution of graphite in cast-iron, aggregates in concrete, fibers in composite material, etc.). In this case the non-destructive signature of each part is sought (ultrasonic celerity, attenuation, number of bursts in acoustic emission, thermograms, etc.) and this signature is compared to that from known standards.

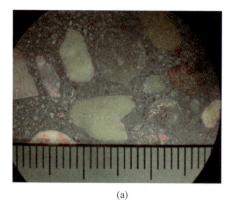 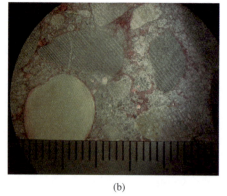

(a)   (b)

Fig. 2.7 Initial concrete (a) and micro-cracking (b) at the aggregate-matrix interface (LCND)

- The thickness of a part, layer, or coating. Some non-destructive techniques are influenced by a thickness law (attenuation, time of propagation of a wave, air-gap etc.) and are completely operational in these metrology measures (Fig. 2.8).

END measures can be carried out following the relative evolution of a physical parameter, as with defect searching, but in more complex cases where there is no zero point, an absolute measurement is often necessary.

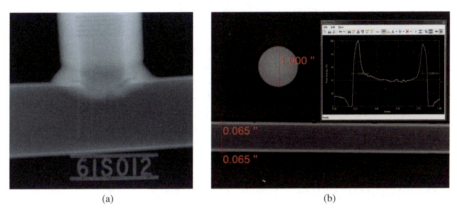

(a)   (b)

Fig. 2.8 Tube tangential radiography (a) and adequate IQI (CIE), radiography and tube thickness (b) (VIDISCO)

2.3 CHOOSING THE OPTIMAL NDT METHOD

Studying the different methods of non-destructive testing shows that when a search for defects is to be considered, the choice of the proper technique is entirely linked to the control of the following characteristics:
- What is the desired analysis between (1) detection, (2) location, (3) characterization, and/or (4) dimensioning of a defect?
  A good analysis will help to best define the answer; as the 4 issues are quite different, there can theoretically be 4 different optimal NDT methods.
- What is the nature of the part and its materials? What are the dimensions?
  Physical characteristics about electrical conductivity, magnetism, and dimensions are of major importance because this will eliminate some NDT methods from the choice.
- What is the nature of the expected defects? What sizes and orientations are being searched for? Where is their presumed location?
  The notion of desired size for the smallest detectable defect is an old one. Today, it is better to talk about minimal (detectable) size of the (maximal) tolerable defect, as this approach is closer to a logic of design that is more tolerant with regard to defects. The improvement in dimensioning possibilities, combined with a better understanding of defect growth mechanisms (serviceability, critical size, etc.) was the starting point for this NDT evolution.

- What are the environmental constraints and thus the consequences for implementation: easy access to the part, and to all the elements of the part, possible or non-desired disassembly, and hostile environmental factors such as temperature, radiation, or an off-shore location? The next chapters will show the potential of NDT methods for solving these problems and the notion of **RC-CND** (*Recommandations de conception des structures basées sur les règles de CND* – Recommendations for structure design based on NDT rules) will also be introduced. This is an approach that consists of upstream planning in structure design of zones that will make the implementation of the presumed NDT method easier.
- Is it possible to have indisputable traces (evidence) of the testing? And in the same logic, can these traces be kept over a certain period of time?
- Is automation possible? This is not the case for all methods.
- What is a possible definition for a standard part and a standard defect that can show and prove that the most appropriate method has been selected, one that detects the minimal defect when it is present in a part, and, of course, does not detect it when it is absent from the part (notion of pseudo-defect)?

2.4   QUALIFICATION OF NDT METHODS

Generally, the **qualification** of an NDT process is obtained when, after agreement between the different parties over what constitutes a standard part with standard defects, the testing finds (and/or characterizes, etc.) without ambiguity (no subjective interpretation from the test operator) the pre-defined minimal defect.

This simple principle immediately raises essential questions:

- How should the standard part be as compared to the real part? Does it have to be strictly identical in shape and material? Can another material be used (a cheaper one)? Another dimension (cheaper, more easily transportable, etc.)? And in this case, is the reproducibility of the heterogeneities, anisotropies respected? And does it have to be respected?
- How should the standard defect be as compared to the real found defect? Can it be artificial (machined by electro-erosion, for example) or does it have to be necessarily real? In other words, how is it possible to use an artificial defect, always less expensive to make, which is representative of the real defect?
- How to manufacture this part? This defect? What are the machining or special methods which will help in reaching it (Fig. 2.9)?

We shall see that each NDT method has specific needs that depend on the laws of physics they use. For example, ultrasonic methods, based on wave reflection, only need a surface and are easily employed to simulate the presence of a crack, but electromagnetic methods, requiring the real volume of the defect to be taken into account, are not satisfactory for this type of approximation.

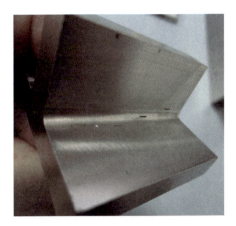

Fig. 2.9 Standard parts with electro-eroded defects (LCND)

The feasibility of the standard part and the standard defect is another important criterion in the choice of the correct NDT method. This is one of the major reasons why ultrasound testing is currently the most commonly used method, as well as the most studied, when trying to solve highly complex NDT problems.

2.5 TESTABILITY OF MATERIALS AND STRUCTURES

This paragraph shows through a few representative examples how far the knowledge of a complex material (or structure) can (and must) go to understand the difficulties in the proposed non-destructive testing and how to overcome them. This leads to the definition of an additional property of materials and structures: their testability.

(Non-destructive) *testability* can be defined as the "ease" with which a material and/or a structure can be examined in a non-destructive way to produce a reliable diagnosis in the detection of defects and/or characterization of those defects (identification, location, dimension) and/or characterization of properties (mechanical, electrical, geometrical etc.). The notion of "ease" is, of course, a subjective one, and one of the goals of this book is to clarify its meaning.

This paragraph is completely related to the notion of RC-CND presented in chapter 14. Defined by the LCND, these rules of design that take NDT into account, consist in anticipating (upstream, at the moment the structures are being designed or even during the development of the materials) the NDT or NDE problems to come. And to proceed so that the production of materials or the design of structures makes them more easily testable.

2.6   MATERIALS AND TESTABILITY:
THE EXAMPLE OF STRUCTURAL MATERIALS

In the domain of NDT, the phrase *structural material* is used to describe a real material whose reaction to testing is not equivalent to that of a perfect material.

For example, in the case of ultrasounds, the perfect medium is elastic, linear, homogeneous, and isotropic. Ultrasonic propagation corresponds to the theory that there is no attenuation by absorption, no deviation, and no echo other than those due to the presence of the defect or the geometry of the part. In radiographic testing, the perfect material has to be homogenous in density. In the same way, the variations in magnetic permeability create difficulties when testing with eddy currents.

When a material is not perfect, there can be attenuation and "structural noise" expressed as interference signals, all of which contributes to a weakening in the power of detection. The signal-to-noise ratio (SNR) decreases and the testability of the material becomes poor, or even non-existent in certain cases. The following chapters will introduce issues of SNR for each method, and chapters 13 and 14 will detail degradations and consequences more specifically, as well as possible solutions.

2.6.1   Example of a problematic: NDT of large thickness
stainless austenitic steel, welded or not welded

Stainless steel is steel that contains at least 12% chromium. Depending on the introduced additional elements, 3 different structures can be obtained: a ferritic stainless steel (of X8Cr17 type, which contains more chromium, an alpha-phase element), a martensitic stainless steel (of X30Cr12 type, which contains more carbon), or an *austenitic stainless steel* (of X2CrNi18-09 type, which contains more nickel, a gamma-phase element).

Austenitic stainless steel is used in the nuclear industry, especially for its properties in high temperature strength and resistance to corrosion. This material is used in many components, the thickness and access characteristics of which necessarily lead to ultrasonic NDT. But as at ambient temperature it retains the austenitic structure of the high temperatures (see iron-carbon diagram, Fig. 2.19), it solidifies with a rather large *grain size*. This solidification problem, that occurs in casting, can also be found in welding, as the growth of the deposited metal takes place through *epitaxy* and globally retains the size of the initial grains. In addition to high work-hardening rates that can be applied to this ductile material by forging, all these production operations generate *morphological textures* (the shape of the grains) or *crystallographic textures* (statistics of crystalline orientation) capable of perturbing ultrasonic propagation.

A multi-pass austenitic weld is anisotropic and heterogeneous because of the elongated grains developing along the heat dissipation lines, in a privileged crystallographic direction (<100>), that grow by epitaxy, after successive welding passes. This texture causes an anisotropy, which in the case of ultrasonic testing will cause variations of speed, attenuation, and deviation. If there is heterogeneity, there will be deformation of the wavefronts, the beam curve, and/or the generation of secondary waves.

It is possible to predict the propagation of ultrasonic waves in austenitic welds by considering them as orthotropic, that is, corresponding to an orthorhombic system with 9 independent elasticity constants. The grain sizes, always an important element in this kind of welding (Fig. 2.10a), can be bigger than the wavelength and generate a reflection-transmission phenomenon at grain boundaries. These multiple reflections produce interference echoes on the signal (structural noise) and an important attenuation that degrades the signal-to-noise ratio.

(a)   (b)

Fig. 2.10 Micrographs of a weld (a) and of a molded part (b), in austenitic stainless steel, showing different textures (EDF)

Molded austenoferritic steel (Fig. 2.10b) is used in the production of primary system components for pressurized water reactors. After solidification, there are two kinds of structures:

$$\text{type A for } \frac{\text{equivalent chromium\%}}{\text{equivalent nickel\%}} > 2$$

$$\text{and type B for } \frac{\text{equivalent chromium\%}}{\text{equivalent nickel\%}} < 2.$$

In type A structures, attenuation is strong but homogeneous. Behavior is globally isotropic and deviation is weak. Variations in speed and attenuation remain acceptable, and it is possible to predict the ultrasonic reaction. In type B structures, attenuation is strong and changes with the movement of the transducer. The material behavior is strongly anisotropic, and there can be an important deviation, heterogeneous in depth. Variations in speed and ultrasonic attenuation are very high, and no prediction is possible.

Ultrasound testing of type A structures is difficult but possible, while that of type B structures is very complex, nearing the bounds of possibility.

2.6.2   Example of a problem: NDT of the waviness of plies in composite materials of CFRP type

In polymer matrix composites, the main mechanisms of damage and rupture modes are identified: cracks in the matrix, matrix/fiber separation, fiber rupture, and delamination (Fig. 2.11a). Except for external aggressions such as impacts, these modes of rupture appear from pre-existing defects in the composite. The pre-existence of delamination, fiber rupture, or cracks in the matrix can be linked with the production process, particularly during the baking phases when heterogeneities in temperature may appear.

In a composite material, the fiber orientation is selected according to the expected constraints and the mechanical characteristics of the parts. The orientation of each ply occurs during the lay-up operation, which means before the shaping of the part. During shaping, fibers may move, which locally modifies the orientation and/or the distribution of the reinforcement in the part. Fiber displacement is thus an implementation defect strongly related to the production process.

The origins of these fiber displacements are identified: residual constraints that may exist in laminate (especially those linked to temperature variations); geometric variations of the plies or of tooling during baking phases, or when cooling, due to the differences of expansion coefficients between the fibers, the matrix, and the molds; and the pressure of the plies stacked one on top of the other in the cases of particular geometry (interlaminar constraints).

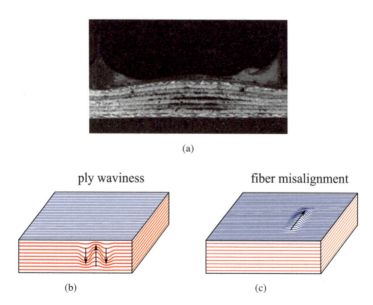

(a)

ply waviness   fiber misalignment

(b)   (c)

Fig. 2.11 Micro-delamination (a) (DASSAULT AVIATION) and fiber displacement defects (b, c); interfaces between the plies are represented in red, fibers are represented in blue.

Two kinds of displacement are identified: ply waviness and fiber misalignment (Fig. 2.11b and c). Ply waviness is a local fiber displacement within the material thickness. Fiber misalignment is a local displacement of fibers in the plane of plies.

Ply waviness and fiber misalignment share the same origin but have different repercussions. The most feared defect is ply waviness; due primarily to its strong presence in thick composites (over 8 mm), this defect heavily degrades the mechanical properties of a part:

- By reducing the mechanical constants, mainly in the longitudinal direction of the undulated fibers. The longitudinal elasticity modulus of fiber dramatically drops (50–80%) with the evolution of the waviness parameter (the height to width ratio of the waviness).
- By reducing the rupture strength. Under stress, the ply waviness itself generates the apparition of damage, such as delamination or intralaminar cracks.

Works in the LCND have shown that it is now possible to detect ply waviness by studying the deviation of an ultrasonic beam.

2.6.3 Example of a problem:
NDT of concrete degradations

Concrete in certain works can have defects making it incompatible with the imposed constraints. These defects, of mechanical or chemical-mechanical origins, appear during concrete solidification or after an evolution in time (Table 2.12).

When macroscopic they mostly influence work resistance, and when microscopic they affect concrete resistance itself. This is then called concrete damaging. It has been suggested that up to 50% of engineering structure works are "sick" in some industrialized countries.

Generally, the NDT methods used for concrete are the same as those used for other structure materials. But the specific problems are more complex: heterogeneity of the concrete composition, the iron framework problem, the type, the rate, and the evolution of the degradations, the size of the searched defects with respect to dimensions, the influence of the constraints, the non-representativeness of the test specimens, etc.

Concrete also presents a great number of difficulties that can be found in ultrasound testing, such as strong attenuation, structural noise, beam deviation, etc. Quite a lot of parameters have an influence over the value of the ultrasonic celerity and some of these parameters are common to metallic or composite materials, others are specific to concrete.

Table 2.12 Various degradations found in concrete

DEGRADATION	CAUSE	SYMPTOM
Chemical attack	Alkali-silica reaction	Expansion then destruction by cracks
	Attack on cement by sulfates	Expansion then destruction by cracks
	Efflorescence and filtration	Porosity and permeability
	Acid attack on basic cement	Erosion
	Salt crystallization in concrete pores	High pressures and cracks
Physical attack	Humidity variations = volume variations	If impossibility to vary, creation of cracks
	Frost-thaw cycles	Cracks, disintegration, and/or separation
	Temperatures cycles	Drop in mechanical characteristics, cracks and separation
	Irradiation by neutrons or gamma rays	Drop in mechanical characteristics, in the Young modulus, cracks and separation
	Abrasion, erosion, cavitation	Loss of material
	Fatigue, vibration	Micro-cracks then large cracks
	Swelling (creep)	Cement cracks, deformations
If iron framework	Problems if 1) corrosion 2) high temperature 3) irradiation 4) fatigue	1) Cracks, fragmentation 2) 3) and 4) degradation of mechanical performance
If pre-constrained reinforcement bars	Problems if 1) corrosion 2) high temperature 3) irradiation 4) loss of pre-constraint	1) section reduction, cracks 2) and 3) decrease in mechanical performance 4) deformations and cracks

2.6.4   Example of a problem: NDT of processed-food materials such as the characterization of foie gras

The non-destructive testing of food-processing industry products has to meet particular requirements, and especially that of the non-contamination of the tested products. Rapid testing speed is also frequently required, justified by high production rates. Industrial foie gras is a particular "structural material," but it is a good example to once again illustrate the level of knowledge about a material that is sometimes necessary in order to deal with the issue of choosing an optimal NDT method.

The name "bloc de foie gras avec morceaux" (foie gras block with bits) is defined by French state decree as "a preparation composed of foie gras with reconstituted bits and seasoning." The total weight of the bits must represent, according to the standards, between 30 and 50% of the finished product, the remaining weight being emulsion. The non-destructive evaluation in respect of the standard, that is, of the proportion of foie gras bits in the blocks and slices, under blister or in cans, before or after pasteurizing, is a particularly complex NDT problem, and is a problem on which the LCND laboratory has worked.

The physical and acoustic characteristics of foie gras (see ultrasonic speeds Fig. 2.13) must be studied to check if the material remains stable (in terms of bit-emulsion percentage) between the pasteurizing phases and under the action of ultrasonic vibrations. Liver density is very close to that of other tissues. Fat can be distinguished from other tissues by its lower density.

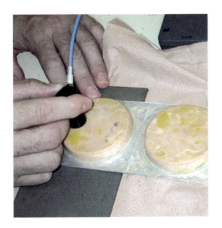

Material	Pressure wave speed (m/s)
Foie gras, raw emulsion	1,610
Foie gras, raw bits	1,666
Foie gras, cooked emulsion	1,564
Foie gras, cooked bits	1,616
Human liver, in vivo	1,540
Human liver, postmortem	1,569
Beef liver	1,566
Pork liver	1,582
Rabbit liver	1,607

Fig. 2.13 Foie gras testing (LCND). A few ultrasonic speeds (at 1.5 MHz) in mammal livers.

Higher quality foies gras do not present macroscopic lesions. Nor are there hemorrhages or necrosis zones. The lipidic melting results from liver lipid exudation caused by heat during heat processing. The exudate lipids solidify around foie gras in a yellow or whitish layer, slightly granular at ambient temperature. The state of this fat may change from liquid to granular at ambient temperatures.

This type of knowledge, particularly used in medical echography, indicates that choosing ultrasounds for industrial foie gras testing may not be the best choice. The texture of the lipidic layer has an effect on the propagation of the acoustic waves. The magnitudes of attenuation and its frequency dependence are mostly the same for fat and for sound liver while differing between liver and lipidic film. Shear stresses (viscosimetry) at kHz frequencies modify the structure of fats and even the critical temperature for the passage to emulsion changes according to the frequency.

2.6.5   Influence of testability on the choice of structural material

If material testability (for an NDT method) was considered upstream of material design, that is, at the stage of steel procurement for metallic materials, for example, this would bring precision to material quality and make it testable, which means that its properties (mechanical, physical, etc.) for the different NDT methods make their implementation and the reproducibility of the measurements possible.

Here are three examples of material choice to help understand this approach:

- The execution of welds in austenitic stainless steel according to a **WPS *(welding procedure specification)***, which takes into account the sequence order in which welding passes are executed, would help reach the desired anisotropy, making it easier to test. The LCND laboratory has been heavily involved in this case, developing the MINA model (chapter 14) for the prediction of crystalline growth with austenitic stainless steel multi-pass welds (Fig. 2.14), which helps understand the propagation of an ultrasonic beam.

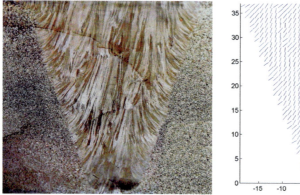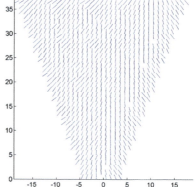

Fig. 2.14 Austenitic steel multi-pass welds (40 mm thick, 30 passes), and prediction by MINA modeling to a lower scale than the pass (LCND, EDF)

- The use of 100% austenitic or 100% ferritic stainless steel tubes enables more efficient eddy current testing. In the case of austenoferritic steel, the structural noise generated by the two types of grains, magnetic and non-magnetic, can be so great that the detection of a small defect is not guaranteed.
- The manufacturing of bolts with ultrasonic quality, that is, having between them a small variation in acoustoelastic properties would help in the proposition of a high-performance ultrasonic method for checking tightening strength.

2.7 MANUFACTURING PROCESS AND TESTABILITY: THE EXAMPLE OF WELDING

As already indicated in the previous paragraph, welding does not only consist in "filling" a chamfer and thus "automatically" ensuring a good junction between two parts. This is a complex technology, and precise knowledge in welding operations helps predict possible defects, their nature, orientation, etc. in order to select one or several NDT methods that perform well.

2.7.1 The different welding processes

These processes differ (Fig. 2.6) according to the source of energy (electrical, thermal, mechanical), and to the protection of the welded junction (flux, gas, vacuum). The main processes are **shielded metal arc welding (SMAW)**, a classic method, which also enables complex welds to be executed, **gas metal arc welding (GMAW)**, which improves productivity (in a flat position), and **gas tungsten arc welding (GTAW)** used for a smaller thickness and appearance, etc.

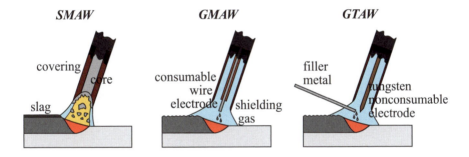

Fig. 2.15 The three classic welding processes

2.7.2 Arc welding with covered electrode (CE)

We are going to outline this process in detail, because it is the most commonly used process today. The general operation is simple, an electrical arc is produced between an anode and a cathode (Fig. 2.16):

Fig. 2.16 The arc is linked to the cathode. The anode has a higher temperature.

From this, there are two possible consequences that may cause defects: the arc is linked to the cathode and the anode is at a higher temperature than that of the cathode. A choice must be made to use either the part or the electrode as a cathode. And there is sometimes an (apparent) incompatibility when one wishes to control both the arc (position in ceiling, for example) and use basic electrodes, which guarantee the mechanical properties but requires "a little help" to melt. The solution consists in using the electrode as the anode and to have the weld executed by a qualified professional.

Physically, it is also possible that arc welding may cause defects (like porosities and/or oxide inclusions) when the metal in fusion is no longer protected by plasma and/or surrounding gas. Figure 2.17 shows both cases, when there is an arc deviation (flexible electrical conductor) from the plasma (gas).

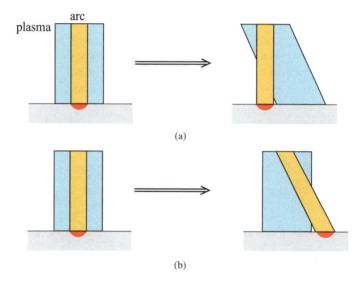

Fig. 2.17 The effect of a draught (a) and magnetic effect (due to a nearby mass for example) (b)

Roles of the electrode's core and covering

The core conducts the current and constitutes most of the filler metal. The covering has several functions: electrical (arc striking, preservation, and stability), mechanical (division of the molten metal, shape of the bead), and metallurgical (preventing over-rapid cooling, protecting the hot metal, making degassing easier, oxide slagging, loss compensation). The metallurgical role is mostly held by slag during solidification, which is why it must be left to operate, before removing when it is cold – a compulsory operation.

Choice of covering

The essential required criteria for a weld (stable arc, correct appearance, mechanical properties) are quite often incompatible with today's electrodes. The choice must be then made between an acid or oxidant covering (for appearance), a rutile covering (for ease of operation), or a basic covering (for mechanical properties).

The basic covering is indeed the only covering that guarantees a low rate of deposited hydrogen (providing that the electrode is baked – portable or storage stove). This low rate helps prevent hydrogen embrittlement of the weld, as illustrated by the evolution of resilience K in Figure 2.18a.

This behavior of steel, changing from a ductile to a brittle mode of rupture is the same as when the temperature decreases (Fig. 2.18b). This well-known phenomenon (as occurred in the impact of the Titanic with an iceberg) helps define a ductile to brittle transition temperature in (complex) relation to the operating temperature.

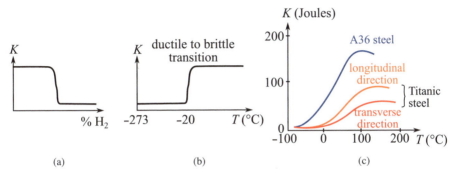

(a)   (b)   (c)

Fig. 2.18 Steel hydrogen embrittlement (a) and cold embrittlement (b), transition temperature for modern steel and Titanic steel (c)

2.7.3   Metallurgic analysis of welding

The thermal cycle sustained by a weld helps illustrate the temperatures reached on heating (Fig. 2.19) via the ***iron-carbon diagram***, which shows the evolution of the working transformation temperatures of unalloyed steels in the temperature function

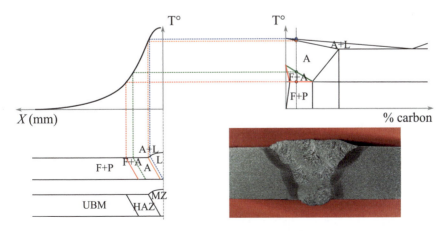

Fig. 2.19 The three metallurgic zones of a weld according to the reached temperatures. The first transformation line (red dots) is called AC1, the following one (green dots) AC3. Macrograph (GROUPE INSTITUT DE SOUDURE)

when heating and cooling speeds are slow (condition of equilibrium, thus of *annealing*). There are 3 zones, the molten zone (MZ) (passing through a liquid state), the *heat affected zone (HAZ)* (passing through an austenitic state), and the unaffected base metal (UBM) (material remains constituted with ferrite and perlite).

Each of these three zones has a weld history and its own potential for defect creation.

Defects in the UBM

In the UBM, there is, by definition, no reason to fear transformation. Temperature rise takes place and expansion-shrinkage constraints are all that remain. If they are minimal (small thickness welds, for example) no defect is to be expected, otherwise there will be geometry deformations, stresses, and only exceptionally, cracks.

Defects in the MZ

In the molten zone, a mix between base metal and filler metal takes place. The chemical composition is predicted according to the dilution rate. The structure solidifies with dendrites (Fig. 2.20), which induces a strong anisotropy and thus possibilities of ultrasonic deviation, for example.

 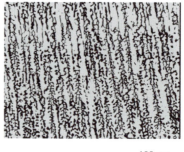

100 μm

Fig. 2.20 Solidified weld structure (dendrites) (EDF)

A greater risk, called hot cracking (see Fig. 2.2), is linked to cracking and takes place at a temperature of around 1000°C especially in the case of a strong clamping of the parts (causing a shrinkage located in the MZ), a presence of sulfur and phosphorus (still liquid at 1000°C), and a large fusion pool, which amplifies the phenomenon. This defect often takes place in the median plane of the weld.

Other possible defects can occur in the MZ: blowholes, inclusions (slag from CE and tungsten from TIG, as their densities are respectively lower and higher than steel, making them appear darker or lighter in radiographic testing), shape defects, incomplete fusion defects between the MZ and HAZ (difficult to see in NDT because even if there is a singular shape, the difference of material is only very slight).

Defects in the HAZ

In the HAZ, all the *heat treatment (HT)* problems can be found, caused by the presence of *austenite*, the only structure likely to undergo a metallurgical transformation. When studying a *continuous cooling transformation CCT* curve (easier to use than the CCTWC, CCT in welding conditions), there are simply 2 cases to be distinguished: quenching or no quenching (Fig. 2.21).

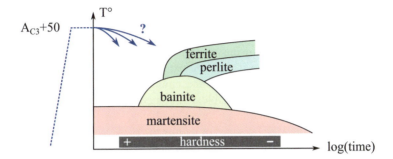

Fig. 2.21 Depending on the cooling, quenching either happens (transformation into martensite) or it doesn't. Note that the CCT curve moves to the right (meaning more "quenchability") if there are additional elements in the steel.

In the case where there is no quenching, this means cooling has been "slow for steel" (an annealing-type cooling; note though that this slowness depends on the type of steel), and the only risk comes from the phenomenon of grain size increase according to the reached temperature (time does not matter here). Below A_{C1} (Fig. 2.19), the grain size doesn't change, between A_{C1} and A_{C3}, the size is reduced, which is positive. Beyond, the grain size increases and as a consequence, resilience K decreases (Fig. 2.22).

In this case, in short, the troubles to come are weak resilience and the presence of internal constraints (from expansion-shrinkage). The solutions consist in accepting this as it is or "relieving the stress" (300°C treatment) to suppress the constraints, or to execute a regenerating annealing to obtain a strong resilience again.

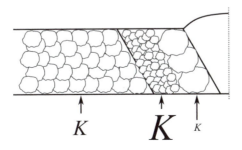

Fig. 2.22 Consequences of grain size increase on resilience K in a weld

When there is quenching, this means cooling has been "fast for steel," thus resulting in martensite (+/– bainite), with a hard and brittle structure. There are two types of internal constraints, those from dilatation-shrinkage and those caused by transformation of austenite to martensite with an increase in volume.

The presence of hydrogen (trapped in the dislocations), strong clamping, and possible metallurgic defects in the part (constraint concentrations) may cause cracks. These cracks appear during the formation of martensite, around 300°C (see previous chapter, Fig. 1.4), and are called "cold cracks" as opposed to the 1000°C "hot cracks."

This cracking can be delayed in time and may appear after 24 hours. Some accidents happen (pipeline fractures, etc.) when the NDT executed right after the execution of welds declares them "compliant" too early. Cracks may indeed appear during these 24 hours via the (simplified) following mechanisms: pipeline in the desert, hot temperature in daytime, cold temperature at night, shrinkage, dislocation movements, migration of trapped hydrogen, increase in the local rate, embrittlement, constraint, and cracking.

In short, when quenching is possible, but there are not any cracks yet, the problems encountered are brittle martensite, weak resilience, expansion-shrinkage, transformation constraints, and the presence of hydrogen. There is a high risk of the structure cracking and then breaking, so it is important to take action (Fig. 2.23).

Fig. 2.23 Brittle fractures in welds, Schenectady (1943) and Flare (1998, CANADIAN ARMED FORCES)

A solution could be to proceed to tempering (300°C, or more if the steel is of a refractory type), which suppresses the constraints and releases hydrogen. Another possible solution is to use annealing, which would help retrieve the material's initial resilience.

But in a generic case where quenching happens, and as it is impossible to know whether a crack is initiated, it is better to avoid this situation entirely by implementing preventive solutions such as pre-heating, which stop the apparition of a quenched structure.

The *weldability* of steel refers to its capacity to resist quenching and not undergo cold cracking.

2.7.4   Conclusions on weld testability

In relation to the detection of defects present in a weld, it is clear that searching (by ultrasounds, eddy currents, radiographic testing, etc.) is easier with prior knowledge of the possible type of defect.

For example, as cold cracks are branch-like, they can be detected via different test angles (in ultrasonic testing) or with different source-part relative positions (in radiographic testing). This is not the case with hot cracks, which are plane. And for the characterization of defects (volume or non-volume identification, 3D location, 3D dimension) the same kind of reasoning is used.

It can be added that the geometry of the weld and its environment also have an influence on the optimal NDT method. Ultrasounds offer the possibility of testing on one face only (radiographic testing does not), but they require enough space to let the transducer(s) move backward when testing under oblique incidences.

2.8   FATIGUE AND TESTABILITY

As the cyclic failure load of a material is smaller than its static failure load, when constraints vary in time, a succession of modifications of the local properties of the material will, after a period of "incubation," cause the initiation of a crack and then a propagation of this crack before the final rupture.

The prediction of the number of cycles before fracture depends on the amplitude of stress, its average value, the surface condition of the part, and the background (temperature, etc.) in which the structure is used. Generally, fracture topography shows two zones: a smooth zone corresponding to the friction of the two faces of the crack while propagating and a granular zone corresponding to the final ductile fracture. Their respective extents depend on the material, notch effect, temperature, corrosion, and constraint variation. The smooth zone helps validate the initial crack zone.

The crack appears in the most stressed zone of the part. This stress value corresponds to the normal operation of the structure and/or the concentration resulting from geometry and/or from present metallurgical defects. But note that a fatigue crack is caused by operational stress and is propagating. This means that when the structure is not in operation, the crack may as well be closed!

In NDT, the aim is to search for fatigue cracks on the surface of parts (emerging cracks) with appropriate technologies (penetrant testing, etc.), but sometimes the crack may not be an emerging one and may result from an internal defect. This is the case for the fracture of this rail (Fig. 2.24), which shows that eddy current testing (rapid, without contact, and adapted to parts of a similar length) wouldn't have been the optimal NDT method as it wouldn't have detected this defect, located out of reach. Only ultrasonic testing is adapted to this kind of deep defect, but unfortunately it takes much more time.

Similarly, one of the means available for improving the fatigue behavior of a part consists in generating compressive stresses on the surface (using shot-peening, for example), which will block the tensile stresses responsible for the future cracking process. These compressive stresses can then disrupt the ultrasonic propagation as they

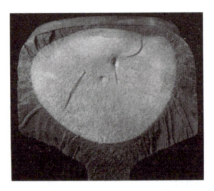

Fig. 2.24 Fatigue fracture within a rail (DUNOD)

have an influence on the surface wave propagation. However, ultrasonic testing is the only "industrial" non-destructive method that enables the evaluation of these constraints.

2.9   CORROSION AND TESTABILITY

Corrosion is a physico-chemical interaction between a metal and its surrounding background; it causes modifications in the metal's properties that may lead to a significant degradation in its function. Corrosion may be a material oxidation at high temperature or a humid corrosion. In the latter case, there are many various forms of corrosion: uniform, galvanic, crevice, pitting, intergranular, by erosion, by abrasion, by cavitation, stress, etc.

The parameters of influence on testability, specific to corrosion damage, mainly concern the location of material loss (or transformation): uniform damage or that localized in the form of pitting, cracks, clogs, blisters, etc. *Stress corrosion (SC)* is particularly dangerous and should be searched for in the zones where residual mechanical constraints are possible, for example, in HAZ of non-relieved welds. *Blistering* is the apparition of blisters or bursting caused by recombined hydrogen having penetrated the steel and caused specific step cracking (Fig. 2.25).

Another example of the necessity of prior knowledge: corrosion may happen when a metallic structure is located in the path of stray voltage. The input surface of a buried metallic structure works as a cathode and the outlet point as an anode. Thus, it is on the latter that corrosion will take place, and this is where NDT will be necessary.

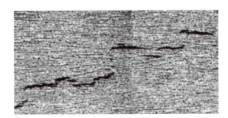

Fig. 2.25 Blistering

2.10 REFERENCES

ASHBY M.F., *Matériaux, microstructure et mise en œuvre*, Editions Dunod, 1985.

ASHBY M., SHERCLIFF H., CEBON D., *Materials, Engineering, Science, Processing and Design*, Elsevier, 2007.

ATTT, *Métallographie et techniques d'analyse*, Editions Dunod, 2004.

BAILON J.P., DORLOT J.M., *Des Matériaux*, Presses Internationales Polytechniques, 2000.

BARRALIS J., MAEDER G., *Précis de Métallurgie*, Editions Nathan, 1994.

BENSIMON R., *Les aciers, tomes I, II, III*, CESTI lectures, Pyc Editions, 1971.

BENSIMON R., *Les matériaux métalliques, Tomes I, II, III*, CESTI lectures, Pyc Editions, 1972.

CALLISTER W., *Science et génie des matériaux*, Editions Modulo, 2002.

CATTANT F., *Materials Ageing in Light Water Reactors*, MAI Lavoisier, 2014.

CHAUSSIN C., HILLY G., *Métallurgie, Tomes 1 et 2*, Editions Dunod, 1974.

CHAKI S., LILLAMAND I., CORNELOUP G., WALASZECK H., Combination of Longitudinal and Transverse Ultrasonic Waves for In Situ Control of the Tightening of Bolts, *Journal of Pressure Vessel Technology*, ASME Transactions, Vol. 129, 2007: 383–390.

CHASSIGNOLE B., *Etude de l'influence de la structure métallurgique des soudures en acier inoxydable austénitique sur le Contrôle Non Destructif par ultrasons*, Doctorate thesis, INSA Lyon, 2000.

CONSTANT A., HENRY G., CHARBONNIER J.C., *Principes de base des traitements thermiques thermomécaniques et thermochimiques des aciers*, Pyc Editions, 1992.

CORNELOUP G., GUEUDRE C., LILLAMAND I., FINE F., Contrôle non destructif du foie gras, Non-Destructive Testing of Foie Gras, COFREND conference, Toulouse, 21–23 May 2008.

DUMONT-FILLON J., Contrôle non destructif, Techniques de l'Ingénieur, ref. R1400 V1, 1996.

DAGALLIER B., MAROY M., *Manuel technique des aciers inoxydables*, Pyc Edition / Semas, 1977.

FRANCOIS D., PINEAU A., ZAOUI A., *Comportement mécanique des matériaux, élasticité et plasticité*, Editions Hermès, 1995.

FRANCOIS D., PINEAU A., ZAOUI A., *Comportement mécanique des matériaux, viscoplasticité, endommagement, mécanique de la rupture, mécanique du contact*, Editions Hermès, 1995.

GAY D., *Matériaux composites*, Editions Hermès, 1997.

GARNIER V., *Evaluation Non Destructive du béton, Contribution des méthodes acoustiques linéaires et non linéaires. Apport de la fusion de données*, HDR Aix-Marseille Université, 2010.

GRONG O., *Metallurgical Modelling of Welding*, Bhadeshia Cambridge, 1997.

HALMSHAW R., *Non-Destructive Testing*, Second Edition, Edward Arnold, 1991.

LAMBERT N., GREDAY T., et al., *De Ferri Metallographia IV, Part I et II*, European Commission, Editions Verlag, Stahleisen, Dortmund, 1983.

LEMAITRE J., CHABOCHE J.L., *Mécanique des matériaux solides*, Editions Dunod, 1996.

LIEURADE H.P., *La pratique des essais de fatigue*, Pyc Editions, 1982.

Mc GONNAGLEW. J., *Essais non destructifs: métaux et matériaux divers*, Editions Eyrolles, 1967.

MOYSAN J., CORNELOUP G., GUEUDRE C., PLOIX M.A., Modeling Welded Material for Ultrasonic Testing Using MINA: Theory and Applications, *Review of Progress in Quantitative NDE*, University of Vermont, Burlington, Vermont, July 17 – 22, 2011.

PAYAN C., GARNIER V., MOYSAN J., Potential of Nonlinear Ultrasonic Indicators for Non-Destructive Testing of Concrete, *Advances in Civil Engineering* 2010, 238472.

POMEY G., RABBE P., *Ruptures de fatigue de pièces de machines*, Editions Dunod, 1968.

PHILIBERT J., VIGNES A., BRECHET Y., COMBRADE P., *Métallurgie: du minerai au matériau*, Editions Dunod, 2013.

RECHO N., *Rupture par fissuration des structures*, Editions Hermès, 1995.

REYNE M., *Technologie des composites*, Editions Hermès, 1995.

SAF, *Guide de l'utilisateur du soudage manuel*, Editions SAF, 1970.

VALLINI A., *Joints soudés, contrôle, métallurgie, résistance*, Editions Dunod, 1968.

VARISELLAZ R., *Soudage, éléments de conception et de réalisation*, Editions Dunod, 1982.

VITTONE R., *Bâtir. Manuel de construction*, Presses polytechniques et universitaires romandes, Lausanne, 1996: 707–733.

XIANG Y. et al., Creep Degradation Characterization of Titanium Alloy Using Nonlinear Ultrasonic Technique, *NDT&E International* 72, 2015.

ZARDAN J.P., GUEUDRE C., CORNELOUP G., Study of Induced Ultrasonic Deviation for the Detection of Ply Waviness in Carbon Fibre Reinforced Polymer, *NDT and E International*, 2013.

CHAPTER 3

VISUAL TESTING

The purpose of visual testing is to identify discontinuities and defects that emerge at the surface. This technique is often used on its own or prior to turning to other NDT methods, because it is simple and cheap. However, the eye can be deceived by different external parameters such as lighting or distance.

Visual inspection is ancient. It belongs to the methods founded on the five human senses, used for millennia to check manufactured objects and products: sense of hearing (tapping – a crack in an amphora changes its resonance frequency), sense of touch (for surface condition, jewel and fabric roughness), sense of taste (for wine), sense of smell (for a perfume), and sight. It should be noted that taste (and smell?) take away some material and so is not strictly a "non-destructive" method.

Visual inspection developed over the centuries with complementary systems and has resulted in two large families:

- Direct visual inspection, by eyesight, supplemented if possible by magnifying means such as magnifiers, microscopes, etc.
- Indirect visual inspection, where the eye inspects images transmitted by instruments like endoscopes, fiberscopes, video-endoscopes, or cameras, some with remote control for hostile environment situations

Visual inspection is used in many cases, for example:

- Checking if a part or an assembly conforms to the part drawing
- Detecting operational anomalies
- Examining parts after operation: wear, corrosion, erosion etc.
- Confirming the existence or the extent of a defect detected through another automatic method

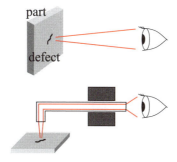

Fig. 3.1 Direct visual inspection or via an endoscope

3.1   PHYSICAL PRINCIPLES

3.1.1   The eye, direct vision

The different types of rays are characterized by their wavelength and energy (Fig. 3.2).
The eye is the receiver of visible electromagnetic rays.

The *spectral sensitivity* of the eye corresponds to wavelengths that lie between 400
and 700 nm (maximum sensitivity is for a wavelength of 555 nm), so green-yellow is
the color we best perceive.

Luminance is a measurable value corresponding to the visual sensation of the
luminosity or brightness of a surface. It is defined by the ray intensity coming from a
source from a given direction, divided by the apparent area of this source in the same
direction. The unit is the candela per square meter $cd \cdot m^{-2}$.

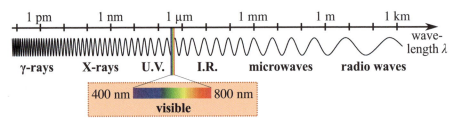

Fig. 3.2 Visible light

Colors are only perceived in photopic luminance (daylight vision) from 1
to 10,000 $cd \cdot m^{-2}$. Vision below this value, from 0.001 to 1 $cd \cdot m^{-2}$, is known as
mesopic (crepuscular), while scotopic vision (night vision) is for a luminance under
0.001 $cd \cdot m^{-2}$. Over 10,000 $cd \cdot m^{-2}$ is a dazzling experience for the eye.

The ability to see two points that are very close together as distinct is called the
spatial resolution of the eye. The maximum angular spatial resolution is 0.003 radian
which means that at a distance of 1 meter, the eye can perceive details of 0.3 mm and
at 10 m, the size of the perceived detail is 3 mm.

3.1.2   The lights used

Illuminance or *luminous emittance* (in lux) is the value defined by the photometry
corresponding to the human experience of lighting. Illuminance is different from energetic emittance through the application of a wavelength weighting corresponding to
the sensitivity of human sight (notion of "reference observer").

The best source of visible white light is daylight. However, it is necessary for
this light to be no less than 500 lux for any local visual inspection and more than
160 lux for any general inspection (European standard EN-13018). This value must
be checked at the surface of the part under inspection conditions.

Color temperature is a characterization of the lighting color. It is measured in kelvin (K). A black body, which traps light and can emit none, is used to qualify the color of a continuous spectrum light. When it is heated, the color and intensity of the light it radiates depend on the temperature. At 500K there is no visible emission. At 2800K, a black body emits a red color visible light (Fig. 3.3). From 5 to 6000K, the light becomes white then blue beyond these values. Thus, the more the kelvin temperature of an object increases, the more it emits a "cold" light for the eye (bluish hue) and conversely. In the collective imagination, the light of a candle is associated with a warm hue (yellow, red) though its actual temperature is low.

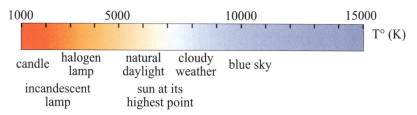

Fig. 3.3 Shading of color temperatures in K, referenced from heating a black body, with examples of corresponding light sources

This temperature scale is used to compare different forms of lighting and understand which to choose. The current use of LEDs, with a convenient reduction of the weight of equipment, enables us to work with temperatures around 6000K. It has been observed that corrosion or cracking is better detected when lit by this white light.

For light sources with a non-continuous spectrum (line spectrum), the ***color rendering index*** (CRI) helps evaluate the quality of light. This is the case of fluorescent tubes and LEDs. The color rendering index of outside natural light, set to 100, is used as a standard to measure all other sources of light. A high CRI indicates that a lamp offers quite a good simulation of the continuous source for this color temperature.

The characteristics of visual inspection are also used in the mode of operation of NDT methods like penetrant testing and magnetic testing. In these cases, white light is used, as is UV light for developers, or fluorescent or phosphorescent tracers.

3.1.3 Magnification with a microscope

An optical microscope comprises two lenses that provide a magnified image of the sample to observe. The object to be observed is placed in front of a first lens, called the objective lens. If it is beyond the focal distance, the inverted real image has a different size and can be bigger than the object. The second lens is ocular: it is positioned so that the image is in its focal plane. In this way, the eye can see an image "ad infinitum."

The ***resolution*** of a microscope (Fig. 3.4) is its ability to separate very close details, which are fundamentally limited by light diffraction. Because of diffraction, the image of a point is not a point but a blur, and the image of two points separated by a distance Δx, that is, distinct but close, will be two blurs that may overlap, thus preventing the distinction of one from the other: the details are no longer separated.

The limit of resolution for a classic photonic microscope is about $0.2\,\mu m$. A transmission electron microscope can reach a limit 100 times smaller.

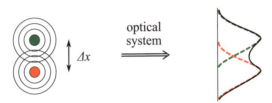

Fig. 3.4 Optical resolution

3.1.4   Indirect vision by endoscopy

Endoscopy is used for observation in zones that are not directly accessible for the eye, through very small openings or even through complex routings. An endoscope can be rigid (Fig. 3.5) or flexible. The image is carried via stacks of lenses or via optical fibers and is seen through an ocular or collected by a CCD or a CMOS and displayed on a screen according to the technology used in the endoscopic equipment.

Fig. 3.5 Principle of a (rigid) endoscope

3.2   TECHNOLOGY AND EQUIPMENT

3.2.1   Sources of light

White light can be obtained by:
- Daylight, the best quality light, though it can only be used for direct visual inspection
- Light generators such as those with a classic incandescent lamp, a halogen incandescent lamp, gas-discharge lamp, or xenon lamp
- LEDs

UV light can be obtained by:
- A mercury-vapor discharge lamp (Wood's lamp)
- Luminescent tubes (neon), possibly

It is advised not to use a non UV-specific endoscope with UV light because of the risks of damage for eye and equipment.

3.2.2 Endoscopes

Industrial endoscopy can be a solution to different situations such as:
- Access to the observation zone (straight or sinuous)
- Access diameter for the passage of the endoscope
- Useful length to reach the zone to be inspected

There are three major families of endoscopes.

Rigid endoscopes, also called ***borescopes***, are favored for applications where access to the inspection zone is in straight line (Fig. 3.6.a). Light and easy to use, the endoscope offers a high-quality image thanks to transmission by optical lenses, perfectly aligned in the tube, which explains the rigidity. The endoscope is equipped with an optical fiber lighting system.

For diameters bigger than 3 or 4 mm, ***video-endoscopes*** (Fig. 3.6b) are used. These are flexible endoscopes, and their objective lens is equipped with a micro-camera transmitting images. The probe is fitted with an orientation system at the end (a rigid section of 30 or 40 mm) that helps orientation within the inspection zone.

For accesses smaller than 3 or 4 mm, ***fiberscopes*** (Fig. 3.6c) are used. These are endoscopes with a flexible sheath and a short rigid section (7–10 mm). They are adapted to the exploration of cavities with a sinuous access or a complex geometry. They are equipped with two networks of optical fibers, one to transmit the image and the other for lighting.

(a) (b) (c)

Fig. 3.6 Endoscope (STORZ), (b) video-endoscope (IT CONCEPT), (c) fiberscope (INDUSCOPE)

3.2.3   Measurement

The measurement can be executed in different ways:
- Comparison of the obtained image with a standard image when it is possible to lock the endoscope at a given distance with a depth gauge
- Comparison with a particular part shape, a screw, etc. geometrically known, on the condition that parallax errors are controlled
- With a reticle eyepiece used to measure the real size of an object on the condition that the distance to the object corresponds perfectly to the magnifying factor used. Such an eye piece (Fig. 3.7) is only to be found in rigid endoscopes.

 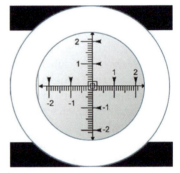

Fig. 3.7 Measurement by reticle eyepiece (OLYMPUS)

3.3   INSPECTION METHODS

Interpreting the indications may seem easy, but it depends on the type:
- Image collection (CCD, CMOS) and transport (lens, fibers, virtual)
- Visualization means (eyepiece, LED screen, plasma or tube, etc.)
- "Warm" or "cold" light (LED, arc, halogen, etc.)
- Angle of light on the objective (annular light or otherwise), angle of vision

The COFREND has issued a guide of good practice (Fig. 3.8) for operators in endoscopy.

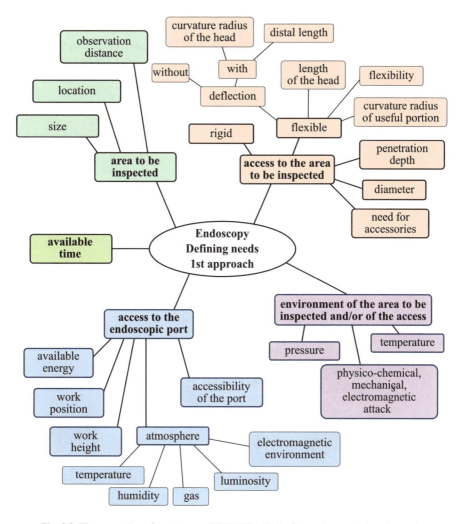

Fig. 3.8 The expression of need (as per COFREND: Guide for good practice in endoscopy)

3.4   EXAMPLES OF APPLICATIONS

Visual inspection, particularly endoscopy, is firstly a preventive inspection of equipment quality, which is carried out without any disassembly intervention, thus helping anticipate non-planned production stops. The replacement of equipment can thus be planned, and the maintenance budget remains under control.

3.4.1   Types of inspected defects

Visual inspection is to be chosen when the geometry of the searched for defect totally or partially stands out on the surface. The following defects can be detected:
- degradation, damage
- cavities, inclusions
- emerging cracks
- corrosion
- steel burns (show better in white light)

3.4.2   Endoscopic inspection

Industrial endoscopy is used for the inspection (Fig. 3.9) of:
- Turbines (visualization of the condition of a turbine's blades)
- Gear reducers or multipliers (visualization of the condition of gears and bearings)
- Combustion engines (combustion chamber: valves, pistons)
- Welds (inspection of the weld quality)
- Tubes (inspection of the internal surface condition)

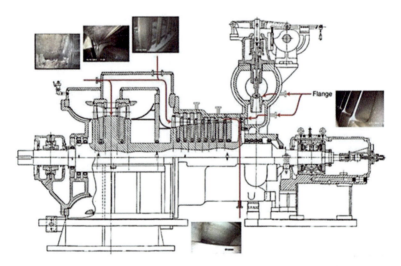

Fig. 3.9 Example of a possible inspection scheme using endoscopy (gas turbine, ATEM)

3.4.3   Inspection of welds

Visual inspection allows us to find emerging defects on the surface of a welded joint (Fig. 3.10). It also gives indications (metal appearance, width of the beads, angle of ripples, etc.) to the observer on the conditions in which the weld was executed and on the probability of internal defects.

Fig. 3.10 Weld visual inspection, with hot cracking (GROUPE INSTITUT DE SOUDURE)

It is an indispensable preliminary to all conventional surface inspections (penetrant testing and magnetic testing) or volume inspection (radiography and ultrasounds). In the case where these are not imposed, it is often visual inspection that helps decide whether a further inspection is required.

3.4.4   Pipe corrosion inspection

The purpose of non-destructive analysis of a pipe network (Fig. 3.11) is to detect calcareous sediment. These represent a disruption in the water flow, and in the case of fire, can mean that the flow is not sufficient to extinguish or control the fire.

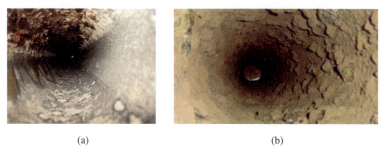

(a)   (b)

Fig. 3.11 Endoscopy of a pipe (a) sprinkler (ANPI) and (b) calcium carbonate (ATEM)

3.4.5   Endoscopic inspection at high temperature

High-temperature endoscopes are equipped with a cooling system using water or air, so that they can work in furnaces nearing a temperature of 2000°C (Fig. 3.12). They are used for quick occasional inspections.

They are used in the steel industry to inspect the wear of refractory material in furnace vaults, glassworks, or in waste incineration to check hearth fouling, etc.

Fig. 3.12 High temperature endoscopy (CESYCO-INDUSCOPE)

3.4.6   Other examples of endoscopic inspection (Fig. 3.13)

aeronautics (OLYMPUS)   wind turbines (INDUSCOPE)

gears (INDUSCOPE)   valves (INDUSCOPE)

landslides or building collapses (SOFRANEL)   corrosion (SOFRANEL)

Fig. 3.13 Different cases for endoscopy

3.5   CURRENT AND FUTURE POTENTIAL

Visual inspection is a high-performance NDT method with a large improvement potential. It has seen progress made in two closely related fields:

- Industrial: in the machine vision (MV) version, which is mainly dedicated to production inspection for large serial processes in many sectors (electronics, food industry, packaging, form recognition, robotics, etc.). MV can check the correct positioning of a part, the absence of a surface defect, whether a bottle or a flask was correctly filled to the correct level, and can carry out form or pattern recognition, etc.
- Medical: endoscopy can be used in the diagnosis or treatment of a disease (surgical endoscopy). Medical endoscopy can be specialized in specific organ inspections. For example, bronchoscopy is the exploration of the bronchial tubes, colonoscopy is that of the colon, cystoscopy the bladder, gastroscopy the esophagus and stomach, and so on. When possible, endoscopies are carried out using natural tracts, otherwise an incision is made (surgical endoscopy).

3.6   REFERENCES

BABB CO, La lumière visible en CND, Babb Co INFO information letter, 2015.

CARON G., MOUADDIB E., *Vision pour la robotique*, Techniques de l'Ingénieur, ref. S7797 V1, 2014.

COFREND, *Opérateurs en endoscopie: bonnes pratiques*, Collection Les Cahiers Techniques de la Cofrend, 2012.

JOLION J.M., Les systèmes de vision, *20ᵉ colloque GRETSI sur Traitement du signal et des Images*, Editions Hermès, 2005.

LA TOISON M., *Eclairage électrique*, Techniques de l'Ingénieur, Ref. D5800 V1, 1993.

VANDENBROUCKE N., *Système de vision industrielle*, Techniques de l'Ingénieur, ref. S7799 V1, 2015.

ZISSIS G., DAMELINCOURT J.J., *Sources de lumière du XXIᵉ siècle*, Techniques de l'Ingénieur, ref. IN26 V1, 2004.

CHAPTER 4

PENETRANT TESTING

This is one of the oldest industrial methods in NDT. Its use can be traced to the 19th century and its rise took place throughout the 1940s and 50s. Penetrant testing completes and extends visual inspection by the detection and location of emerging discontinuities in the surface. The implementation is relatively simple and is conducted in several steps (Fig. 4.1):

- Cleaning the surface to be inspected, scouring if necessary
- Application of a colored liquid called *penetrant* on the surface
- After a waiting period, during which the liquid seeps into the discontinuities by capillary effect, the excess of penetrant is removed by appropriate washing
- After drying, application of a thin powder, the *developer*, through which the penetrant contained in the discontinuities transits, also by capillary action
- Formation of a magnified image of the opening at the surface of the discontinuity
- Inspection, interpretation, and recording of the image, and final cleaning

Penetrant testing is applicable to non-porous surfaces, mostly on metallic alloys. It can be used at different stages of the manufacturing process and in maintenance operations, too. This method offers many advantages and large detection sensitivity. In an ultra-high sensitivity process, it is possible to detect defects of a few tenths of millimeters in length and less than a micron wide.

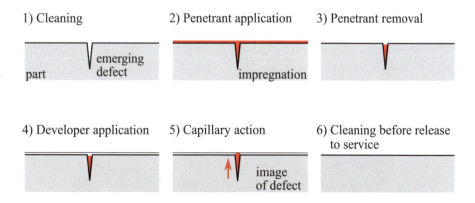

Fig. 4.1 The 6 phases of penetrant testing

This method can be used in a production line or on site and can be implemented in very different conditions of temperature and humidity. It does not require instrumentation.

4.1 PHYSICAL PRINCIPLES

4.1.1 Physico-chemical phenomena involved

Penetrant testing is based on the characteristics of **wettability** and **capillarity** of liquids in contact with solids. These phenomena cause the spreading of penetrant on the surface, the filling of discontinuities, and the upwelling of the same penetrant through the developer.

The spreading ability of a liquid (penetrant) over a solid surface (the tested part) relies both on the internal forces of the liquid and of the attraction exerted by the solid; this is why the distinction is made between how a liquid behaves on its own and how it behaves when spread on a solid.

Surface tension

The interface between a liquid and a gas (the air, for example) is a surface that behaves as if the liquid was surrounded by an elastic membrane. The molecules of the liquid located inside are subjected to cohesion forces, the net force being zero, whereas those located at the surface are mainly attracted towards the inside of the liquid.

This phenomenon is the reason for the coherence of the liquid, at constant temperature and pressure. The force exerted on the liquid surface is called the γ **surface tension** and is measured in newtons/meter. It is to be noted that this value decreases when temperature rises.

Wettability and capillarity

When a liquid is deposited on a solid, the forces of the liquid's internal cohesion compete with the forces of attraction of the solid on the liquid. A liquid can wet the surface of a solid, the static angle α formed between the solid surface and the tangent to the liquid is then under 90°. If the forces of internal cohesion of the liquid are preponderant, the liquid remains in the form of a drop (slightly flattened because of gravity action), and angle α is bigger than 90° (Fig. 4.2).

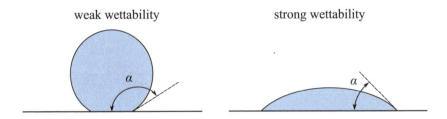

Fig. 4.2 Wettability and contact angle

If a capillary tube (solid) is now dipped into a wetting liquid, the level of the liquid will rise by a height h above the level of the liquid. However, if the liquid is non wetting, the level in the tube will be below the initial level of the liquid.

This phenomenon is illustrated by **_Jurin's law_** (Fig. 4.3), which as a first approach gives the height h reached by the liquid:

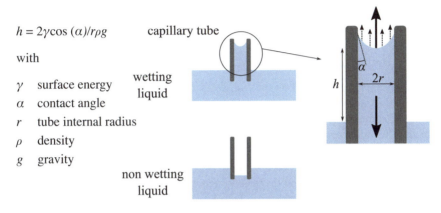

$$h = 2\gamma\cos(\alpha)/r\rho g$$

with

γ surface energy

α contact angle

r tube internal radius

ρ density

g gravity

capillary tube

wetting liquid

non wetting liquid

Fig. 4.3 Jurin's law

A penetrant used in penetrant testing must then have an angle α largely under 90°, as it is this phenomenon that causes it to fill a discontinuity. Note that as a crack is generally closed, it will not be completely filled because of the presence of air that cannot escape. As a consequence, it is impossible to have any access to crack depth information, for example, since there is no control in filling.

This angle α also is the reason for the upwelling of the penetrant in the developer through the interstices between grains that form the tridimensional network of micro-capillaries, causing a magnification of the defect opening.

Penetrant, solvent, and emulsion

A **_solvent_** is a liquid that can dissolve substances in a dilution (not as a chemical attack, however) without chemically modifying them. Solvents are classified according to their composition. Inorganic solvents contain no carbon – water belongs to this category. Organic solvents containing carbon include hydrocarbons, alkanes, benzenes, oxygen solvents, ethanol, acetone, ether and halogen solvents with chlorine or fluorine base, and trichloroethylene.

A classic penetrant is completely miscible in its specific solvent. Conventional solvents, based on oil by-products, are not miscible with water: an emulsion must be created to remove excess penetrant with water.

An emulsion is a mix of thin droplets of one liquid added to another. For stability, a surfactant or an emulsifier is further added. A few pre-emulsified penetrants that can be directly washed with water can be found (see § 4.3.2), and there are a few post-emulsified ones too, containing surfactants. To wash them with water, it is necessary to add an emulsifier to create an emulsion with water.

Penetrant viscosity

The *viscosity* of a liquid is linked to its temperature. This is the ratio between the flow velocity and the constraint creating the flow. In the absence of external constraints, it is the weight of the liquid that causes the flow.

The viscosity of the penetrant must not be too high so that it can spread rapidly to cover the whole surface to be inspected, often the whole part. Another reason that viscosity should not be too high is to avoid excessive consumption (one drop is enough).

4.1.2 Optical phenomena and vision

In most cases the inspection is visual. Detection and observation of indications are obtained from optical phenomenon and particularly in the range of the spectrum of the electromagnetic waves (400 nm to 700 nm) corresponding to the visible spectrum. The characteristics of human vision have a direct influence on the inspection results.

Generally, objects are seen because they emit light or they reflect it. Penetrant agents use either one or the other phenomenon: reflection for colored penetrants, visible in "white" light, and emission based on *fluorescence* for fluorescent products, visible when excited by ultraviolet light, sometimes called "black light."

Vision

The detection of indications from penetrant testing relies on contrast and luminosity phenomena associated with the way an eye works. A typical curve of eye sensitivity (that is, a standard or average) can be established based on the processing of light by the retina and brain. Perception is maximized when there is a contrast in luminosity (black and white) and opposition between the colors. This fact has led to the traditional use of red color penetrants over a white background when the test is carried out in daylight.

Use of white light

When a test is carried out with a dyed penetrant, the white light used contains the whole range of visible radiation. The red penetrant absorbs all the wavelengths except those corresponding to the red color and the white developer reflects all the wavelengths identically.

In order for the retina to be sensitized by the reflected light, this light has to be strong enough, that is, the lighting of the inspected surface has to be strong enough, too. Minimum values of lighting are recommended (standards).

Use of fluorescent light

Fluorescence and phosphorescence are particular cases of luminescence (Fig. 4.4). *Fluorescence* corresponds to a synchronized radiation emitted with excitation, whereas *phosphorescence* can be observed even when the material is no longer illuminated.

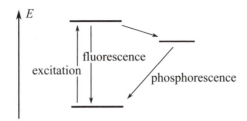

Fig. 4.4 Jablonski Diagram: a molecule releases the absorbed luminous energy rapidly (fluorescence) or slowly (phosphorescence) and returns to its equilibrium state

When a test is carried out with a fluorescent dye, ultraviolet light is used to excite the penetrant, against a dark background. The penetrant (and only the penetrant) then emits a green-yellow light corresponding to a wavelength of 550 nm (maximal sensitivity of the human eye), which contrasts sharply with the surrounding darkness.

The light representing an indication is an emitted radiation and is thus less diffuse than reflected light. This phenomenon associated with optimal contrast helps improve the sensitivity of the test. The resolution power is increased by a factor of 2 or 3 in comparison with a test in white light.

For the retina to be sensitized by the emitted light, this light has to be strong enough. Minimal values of energetic lighting are recommended (standards).

Critical thickness – Beer-Lambert law
When light goes through a liquid, it is attenuated according to the thickness it has to go through. When thickness is small, attenuation is low, and the eye can no longer see the color: the liquid seems to be transparent. This limit is called critical thickness according to the *Beer-Lambert law*.

In the case of penetrant testing for small defects and in order for the dyed penetrant to be as visible as possible, its critical thickness (the limit of detection) has to be as small as possible.

Brightness or luminance
Photometric *luminance* (formerly called *brightness*) characterizes the fact that the eye's sensitivity to the perception of colors varies according to the ambient luminosity. Luminance is a characteristic of fluorescent penetrants. The larger this value is, the more the reading can be done with limited energetic lighting.

4.2   TECHNOLOGY: PRODUCTS AND EQUIPMENT

4.2.1   Cleaners

Penetrant testing cleaners (or stripping agents) are either active products such as highly concentrated acid and alkali detergents or passive products such as solvents. Solvents can be classified according to their composition (inorganic, organic, oxygenated, etc.).

4.2.2   Penetrants

Penetrants are complex products, notably containing a solvent, surfactants, coupling agents and plasticizers, and a dye. There are water-based penetrants too, with very little solvent based on hydrocarbons; however, their use is not yet widespread.

They can be pre-emulsified, which makes them water washable, or they are post-emulsified and then associated with a specific emulsifier to become washable. There are two types of post-emulsified penetrants, lipophilic (water-free) and hydrophilic (with a few tens of percentage of water).

Penetrants visible in white light are generally a "deep" red, with a wavelength of 460 nm. Fluorescent penetrants react to ultraviolet and radiate a "green/yellow" light corresponding to a wavelength of 550 nm.

4.2.3   Removal of the penetrant

Before the phase of developer application, some products (also called cleaners) are used to remove excessive penetrant. Choosing among them depends on the kind of penetrant that has been used, and the surface condition of the part to inspect. Water can be used, notably for *pre-emulsified penetrants*, or an emulsifier diluted in water or oil according to the kind of *post-emulsified penetrant* (hydrophilic or lipophilic), or a solvent with or without water.

This cleaning operation is very important, and it is estimated that it accounts for 90% of the method's success.

4.2.4   Developers and family of products

Developers are mineral powders with a very thin granulometry (a few dozen nanometers), generally based on magnesium silicate, whose high absorption properties promote the capillary action. They are non-fluorescent and include additives to make their application in a thin coating easier.

Developers can be found as a dry powder, the support for their application being then air, or in a volatile solvent, generally alcohol or an aqueous solution. Powders are found in bulk containers and developers with solvents are in sprays. When a colored penetrant is used, dry developers are generally proscribed.

Peel developers that form a film after the support solvent has evaporated also exist. They can be peeled off from the surface and can constitute a replica that can be archived. This technique has long been one of the only means of keeping a record (image) of penetrant testing, but it is now in strong competition with digital photography.

A family of products (Fig. 4.5) is a combination of products that helps comply with the current standards (EN 571) when executing a penetrant test. A penetrant, its removal product, and if possible the developer, should all come from the same manufacturer.

Fig. 4.5 Products can be packaged in bulk in cans or in sprays: for magnetic testing (the 5 on the left) and penetrant testing (the 5 on the right: fluorescent penetrant, red penetrant, white developer, degreasing solvent, final cleaner) (SREM)

4.2.5 Lighting equipment

For colored penetrants, white color lighting can be achieved with daylight, incandescent lamps, diode lamps, or xenon discharge lamps. The inspection should be carried out with a lighting level higher than 500 lux. A luxmeter helps check that the lighting level of the surface to be inspected is sufficient. Its curve of spectral sensitivity must be close to that of the human eye.

For fluorescent penetrants, ultraviolet lighting can be carried out using mercury discharge lamps called *Wood's lamps*, or electroluminescent diodes (LED). The use of neon tubes is generally to be avoided.

The spectrum emitted by these lamps must range between 320 and 380 nanometers of wavelength, the energetic flux must be higher than $10 \text{ W} \cdot \text{m}^{-2}$, and interfering white light must not exceed 20 lux. Ultraviolet radiometers are used to check the level of energetic flux.

4.2.6 Reference parts

Reference parts are used in penetrant testing to check the global capability and reliability of the inspection process implemented on one hand and, on the other hand, to identify its sensitivity. Some of these reference parts are specifically dedicated to periodic evaluation of the performance of the penetrants present in facilities or stored.

Standards (EN 571) recommend different test specimens:

- A nickel-chromium plated part (Fig. 4.6) is used to compare the sensitivity of penetrant families. It comprises a brass plate on which a layer of nickel-chrome (10 to 50 microns) has been deposited and stretched lengthwise to obtain a network of transverse cracks.

Fig. 4.6 NiCr Type 1 penetrant testing check gauge (SREM)

- A penetrant testing check-gauge in stainless steel helps evaluate the global performance of a process by checking the sensitivity and efficiency of the penetrant testing operations. This gauge, 155 × 50 mm, thickness 2.5 mm, for a global check of the colored or fluorescent penetrant testing process, is divided into two zones.

The first, with different types of roughness, is designed to check penetrant washability. The second includes five artificial defects in the form of star-like cracks made by punching on the opposite face and is used to check the sensitivity of detection.

4.2.7 Equipment

Two kinds of equipment can be used in penetrant testing: mobile equipment that can be used for on-site testing or fixed installations. At a minimum, mobile equipment is composed of portable spraying devices for the different products (sprays, for example), cloth, brushes, sources of light according to the process, and personal protection gear.

The material constituting the fixed installations must be able to resist the products used on one hand and, on the other hand, they must not be a source of contamination. The associated devices for ventilation, extraction, and effluent treatment must also be compliant.

Hygiene and safety requirements must be complied with, notably when considering a vapor phase treatment for initial cleaning, which is always complex and involves often highly carcinogenic products.

4.3 TESTING IMPLEMENTATION

4.3.1 Typical process, retest

A process outline describes the chronological sequence of operations according to the products used. Particular protection conditions can be added if necessary, as well as the procedure for a retest. The process outline (Fig. 4.7) specifies the following steps:

- Preparation prior to cleaning, chemical cleaning (remember that 90% of the success of the penetrant testing operation comes from this preparation phase)
- Drying
- Penetrant application method, impregnation time (penetration time)
- Procedures for the removal of the penetrant, checking for the removal of excessive penetrant – the cleaning product is never to be sprayed or splashed (except water), but is instead to be applied using a cloth
- Drying
- Procedures for the application of developer, developing time
- Conditions of testing and observation
- Final cleaning

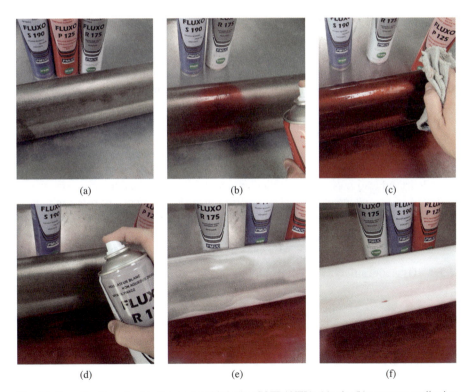

|(a)|(b)|(c)|
|(d)|(e)|(f)|

Fig. 4.7 Example of penetrant testing on a welded tube (SOFRANEL); (a) tube (b) penetrant application (c) removal with a cloth (d-e) developer (f) crack

Retesting is very frequent in NDT and can confirm doubtful cases, notably because the sensitivity used locally this time, will be higher. However, retesting is much more complex to implement in penetrant testing than in other NDT methods, as it requires a special procedure to correctly remove the products already used.

In this case it is possible to use a "super cleaner" (without over-washing) that eliminates all the previous indications, and to check if the zone shows penetrant seeping out again.

4.3.2   Choice of a process, sensitivity, reliability, compatibility

Penetrant testing processes offer different levels of *sensitivity*, defined as the capacity to detect the thinnest defects. Reference parts whose characteristics and number of defects are known are used to measure the sensitivity.

Standards define four levels of sensitivity:
- S1 (low): concerns colored penetrants and some fluorescent penetrants
- S2 (medium): concerns fluorescent penetrants other than post-emulsified ones
- S3 (high): concerns fluorescent penetrants essentially post-emulsified ones
- S4 (ultra-high): for (lipophilic) post-emulsified fluorescent penetrants

Though it is difficult to give a real assessed value to the limit threshold of detectable defects, it is true that ultra-high sensitivity processes can show defects of a few tenths of a millimeter in length and less than a micron wide.

The parameters affecting the sensitivity and quality of a test are mainly the nature of the tested parts, the types of searched-for defects, the mode of application of the products, penetrants, developers, and the environment (lighting, ambient conditions, possibility of using water, etc.). These parameters are not independent and choosing a process is the result of a compromise between the optimal desired sensitivity and actual possibilities for implementation. This is why a test of certain structures might require the use of fluorescent products but the practical impossibility of obtaining a satisfying darkness may result in a change to testing with a colored product.

Penetrant testing is mostly applied to metallic materials, and it is sometimes necessary to take a few precautions. For ferritic and austenitic steel, corrosion inhibitors may have to be used, and for aluminum, zinc or copper alloys (which are sensitive to aqueous corrosion), some particular precautions when rinsing can be useful.

4.3.3   Hygiene, safety, and environment

The use of cleaning products, penetrants, and developers can be damaging to the environment and to the health, for example, penetrants may seep through the epidermis. These products require specific conditions of use, packaging, storage, and recycling, and must be completely destroyed after use. It is thus mandatory to ensure that their safety data sheets available (Fig. 4.8).

Fig. 4.8 Example of safety data sheet (INRS site)

The ultraviolet radiation used to excite fluorescent penetrants belongs to the UV-A range (320 nm to 400 nm) and has a limited effect on the skin at the levels of the energetic lighting used. However, it is advisable for operators to avoid exposing their skin over long periods.

4.3.4 Interpretation and acceptance criteria

During the development phase, the occurrence of a colored or fluorescent trace is a "likely" indication that there is a defect. Inspection is necessary to help separate false indications (interference) from those caused by a real defect.

The only possible information on defect dimension regards the length of the indication. For the rest, some aspects can be taken into account, providing a critical attitude is employed, associated with experience in interpreting situations:

- An indication that appears rapidly is suggestive of an open defect, while a slow indication is most likely to be a narrow one
- An elongated (linear) indication is the sign of a plane discontinuity
- An indication in the form of a smear (non-linear) is the sign of a volumic discontinuity

Acceptance criteria essentially depend on the product standards and on the codes and specific recommendations in a given sector of activity. However, they share a few common points in their different approaches:

- An indication is said to be linear if its length is more than three times its width – these indications can be aligned or independent
- Otherwise, an indication is non-linear and can be aligned or grouped, depending on whether they can be independently identified and form a line or whether they are grouped in a mass

The acceptability or refusal of these indications depends on their dimension, number over a given surface (accumulation), position, the thickness of the part, and on the combination of all these different characteristics.

4.4   EXAMPLES OF APPLICATION

4.4.1   Detectability on rolled and forged parts

At this stage of the elaboration of a product, volume defects or internal point defects can be found. These are caused by the first transformation phase, after drawing or compression, seen as emerging or sometimes linear discontinuities. The detectable defects specific to rolling and forging are inclusion alignments, flaws (metal folds partially re-welded), lines of porosity, and other kinds of folds and cracks if the test concerns the lateral edge of the part with respect to the direction of stress.

4.4.2   Detectability on molded parts

The detectable defects at this stage of elaboration are porosities emerging on the surface (Fig. 4.9), shrinkage cracks, cold shuts, and cavities that were not totally transferred in the deadheads.

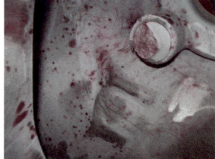

Fig. 4.9 Penetrant testing on an as-cast engine casing and porosity detection (INSAVALOR)

The outlines of *cold drops* correspond to metal drops spurted onto the edges of the mold during the casting stage that solidified (on the surface) before the next cast of liquid metal arrived.

4.4.3 Detectability on machined and treated parts

Detectable defects include scratches, grinding cracks, and quenching cracks. Compared to defects from first and second transformation modes, these defects are much thinner and require a process with increased sensitivity.

4.4.4 Detectability on welded joints

Detectable defects concern welded joints, but also sheet metal: folds in chamfers before welding, transverse cracks, longitudinal cracks (Fig. 4.10), crater cracks, emerging porosities, and troughs in the weld. The latter is generally detected by prior visual inspection, the penetrant testing inspection giving confirmation.

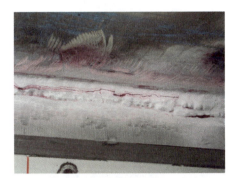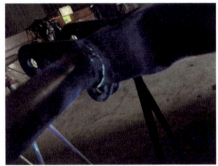

Fig. 4.10 Penetrant testing on welded parts (CNDR and INSAVALOR)

4.4.5 Detectability on parts in service

The main defects detectable after use are fatigue cracks (Fig. 4.11), tears and load cracks, shock cracks, cracks caused by sudden heating and cooling, and flaking. Generally these are thin discontinuities. Note that on-site inspection conditions sometimes make it difficult to apply the ideal testing process.

Fig. 4.11 Penetrant testing on a crank damaged by fatigue (SOFRANEL)

4.4.6   Special applications using particular products

Thixotropic products, penetrants, and emulsifiers are generally not presented under the form of liquids but rather as gels. They are used for vertical or overhanging surfaces. Their use avoids inopportune dripping and waste. The colored ranges specifically dedicated to "high temperatures" allow penetrant testing up to 200°C to be carried out, compared to conventional products limited to temperatures around 80°C or 90°C. The properties of fluorescent products degrade at high temperatures, particularly luminosity, which makes their use inappropriate.

Conversely, the use of conventional products is not to be considered below 5°C because of issues around freezing and the ill-functioning of aerosol spray cans.

When used in electrostatic spraying (Fig. 4.12), penetrants, emulsifiers, and developers reach a high negative potential thanks to an electrode connected to a sprayer. The products are transformed into charged particles that are attracted, as the part is connected to the ground, following the path of the electromagnetic field lines created in the air between the nozzle and the part. They are then deposited in thin and uniform layers on the surface of the part.

Fig. 4.12 Installation of electrostatic (or not) spray penetrant testing (OMIA)

Among the various advantages of this technique are a short application time, a smaller quantity of product used, and the guarantee that the whole surface is uniformly covered. This technique is used for parts with large dimensions and/or with complex geometries, on aeronautic parts such as turbines, for example.

4.4.7   Application to non-metallic materials

Materials of different natures can be tested by penetrant testing, but they must be compatible with the components present in the formula of the different products used in the testing process.

Potentially testable materials are plastics (polymers), composite materials (caution: possible induced coloring), concrete-type organic materials (chemical incompatibility and washing problems are to be taken into account), and ceramic materials (high retention of penetrant). Remember that the viscosity of the products used must be compatible with the (micro) porosity of the materials. Penetrant testing is nevertheless of limited application on non-metallic materials.

4.5 CURRENT AND FUTURE POTENTIAL

Though it is one of the oldest and simplest methods, penetrant testing is still efficient. It is often in competition with magnetic testing. In simple words, penetrant testing is more "refined" than magnetic testing (better sensitivity) but is less "robust" (degradation according to the surface condition).

If either method can be used (a ferromagnetic part, for example), penetrant testing is preferred to magnetic testing when the surface condition is good. Similarly, if a part is either small or very large, penetrant testing will be chosen instead of magnetization, as obtaining a correct result with the latter procedure involves technical difficulties.

On the other hand penetrant testing is complicated and expensive in large series, especially when product recycling requirements are to be met. In this case, magnetic powders from magnetic testing are easier to collect.

The penetrant testing method is still the subject of innovation and improvement.

4.5.1 Product quality, performance, and treatment

The characteristics of penetrant testing products are the subject of continual improvement, particularly the performance of "luminance" and "washability." But in the past few years the most advanced progress have been made in the fields of hygiene, safety, and environment, and improvement is still on-going, for example, in the elimination of trichloroethane, followed by the limitation of trichloroethylene use.

4.5.2 Associated equipment

Lighting devices are continuing to be developed, thanks to progress in semiconductor physics, achieving a better color rendering quality, increased reliability, and a reduction in overall dimensions and weight for the same power.

4.5.3 Observation of excitation and indications

Fluorescent penetrants were initially designed to be excited by an ultraviolet light of 365 nm. The passage to an excitation light occurring in a 405 nm to 450 nm spectrum (visible field: violet/blue) would help solve the problems caused by the risk induced by UV-A to the skin and eyes.

4.5.4 Recording and automatic detection

This is a longstanding improvement objective of this method. The automatic data processing devices used to identify the indications and distinguish them from pseudo-defects remain an important subject for research.

These devices mainly concern fluorescent processes. Problems remain to be solved in data processing in the cases of high background interference, parts with complex shapes, and erroneous luminous indications. Problems in the calibration of standard defects and the introduction of thresholds, necessary for an automatic detection of defects, have to be solved before industrial applications are considered.

4.6 REFERENCES

BETZ C.E., *Principles of Penetrants*, Magnaflux Corporation, 1969.

CETIM, *Traitement des effluents de ressuage*, Editions du CETIM, 2009.

COFREND, *Ressuage: annales officielles de la certification, niveaux 1, 2 et 3*, Editions Lexitis, 2013.

COFREND, *Guide des bonnes pratiques d'éclairage en ressuage et magnétoscopie*, Les Cahiers Techniques de la Cofrend, Editions Lexitis, 2014.

DUBOSC P., CHEMIN P., *Ressuage, Manuel de cours niveaux 1, 2 et 3*, Les Cahiers Techniques de la Cofrend, Editions Lexitis, 2013.

FOREST G., *Choix d'une méthode de contrôle*, Editions Afnor, 1992.

INSAVALOR, *Contrôle Non Destructif par Ressuage niveaux 2 et 3*, Lecture documents, 2014.

TRACY N.A., O'MOORE P., *Nondestructive Testing Handbook, Volume 2, Liquid Penetrant Testing*, ASNT, 1999.

NF EN ISO 3452-1, *Examen par ressuage – Partie 1: Principes généraux*, July 2013.

NF EN ISO 3452-2, *Examen par ressuage – Partie 2: Essais des produits de ressuage*, January 2014.

NF EN ISO 3452-3, *Examen par ressuage – Partie 3: Pièces de référence*, January 2014.

NF EN ISO 3452-4, *Examen par ressuage – Partie 4: Equipements*, February 2001.

NF EN ISO 12706, *Essais Non Destructifs, Contrôle par ressuage, Vocabulaire*, February 2010.

NF EN ISO 3059, *Contrôle par ressuage et contrôle par magnétoscopie, Conditions d'observation*, February 2013.

CHAPTER 5

MAGNETIC TESTING

Alongside penetrant testing, magnetic testing is one of the oldest methods of NDT. Indications of its first applications can be found from the late 19th century.

This method can only be applied to ferromagnetic materials and alloys (based on iron, cobalt, or nickel). It requires a supply of magnetic energy, and in most cases a source of electricity is necessary.

Magnetic testing can be used at different stages of product manufacturing and during maintenance operations, too. The surface conditions and shape of the tested part are rarely an issue. Small or large dimension parts can be tested. The duration of the test for a small part or a limited zone is approximately one minute. It is possible to identify, with high sensitivity, defects of a few tenths of a millimeter long and a width smaller than a micron when the defects are emerging or located very close to the tested surface.

These characteristics mean that it is a method still in use after almost a century of existence, and its robustness of use is of particular interest.

The test comprises the following several steps (Fig. 5.1):

- Cleaning of pieces of grime present on the surface
- Application of a magnetic field leading to the magnetic saturation of the part
- Application of a magnetic developer on the tested surface
- Formation of a "magnetic image" by developer accumulation caused by the leakage flux created by a discontinuity close to the surface
- Inspection, interpretation, and recording of the image
- Cleaning and demagnetization

This implementation is available with different possibilities according to the means and methods of magnetization, the nature of the magnetic developer, the shape of fields used, etc.

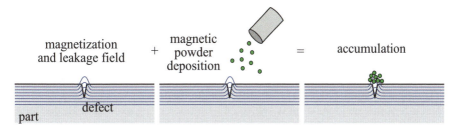

Fig. 5.1 Principle of magnetic testing

5.1   PHYSICAL PRINCIPLES

5.1.1   Electromagnetic phenomena

Magnetic testing is based on the magnetization of the materials and parts to be inspected. When a medium is subjected to ***magnetic excitation H*** (unit: $A \cdot m^{-1}$), generally obtained from a current, it takes on a magnetic condition called ***induction field B***:

$$B = \mu_0 \, \mu_r \, H \quad \text{in Tesla,}$$

where μ_0 is the absolute magnetic permeability of vacuum, constant and equal to $4 \cdot \pi \cdot 10^{-7}$ Henry $\cdot m^{-1}$ and μ_r is the relative magnetic permeability of the medium. The latter is the reaction of the medium to the magnetic excitation, that is, its capacity to modify a magnetic field and thus the magnetic flux lines. Materials are distinguished between ferromagnetic materials (iron, nickel, cobalt, etc.), diamagnetic materials (copper, water, gold, zinc, etc.), and paramagnetic materials (aluminum, magnesium, air, etc.). For a ferromagnetic material, μ_r depends on excitation and can reach large values. Materials that can be inspected by magnetic testing belong to this category.

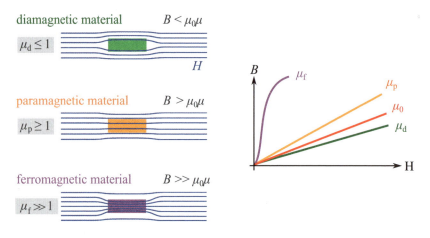

Fig. 5.2 Distribution of field lines according to material permeability

The channeling of the magnetic field in a conductive material is all the more efficient when field variation frequency, permeability, and conductivity are high, resulting from the presence of induced currents.

Ferromagnetic materials: curve of initial magnetization

When a ferromagnetic material is subjected to an increasing excitation H, the curve representing the relation between the induction field B and this excitation H shows several zones (Fig. 5.3 curve from a to b, then c, d, etc.).

The curve (ab), from a to b, is called the initial magnetization curve. It first shows a quasi-linearity between B and H. So, the more the excitation H increases, the more the material reacts. The magnetic state B increases proportionally with H.

The second part of the curve (ab) bends, meaning that the material reacts less to the increase in H. The field B cannot increase as much as H. This is the ***saturation bend***. The third zone shows a very small evolution in B, almost inexistent though the excitation is still increasing. The material no longer reacts to excitation. This is ***saturation***.

Hysteresis phenomena

When the excitation field increases beyond the linearity zone and suddenly stops, its value does not return to zero, it retains a certain value called ***remanent induction Br*** (Fig. 5.3, curve bc). To bring this induction field back to the zero value, an excitation must be applied (and retained) in the direction opposite to that initially applied. This is coercive excitation (curve cd).

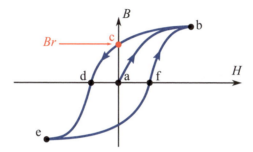

Fig. 5.3 Curve of initial magnetization and hysteresis

By applying a progressive excitation in a given direction, then in the opposite direction, it is possible to draw a closed curve called the hysteresis loop (curve bcdefb). This is the representation of the memory effect of a ferromagnetic material subjected to a magnetic excitation.

To return B to zero without keeping an excitation, a decreasing alternative field must be applied, thus drawing smaller and smaller hysteresis loops and bringing the material back to a neutral magnetic state.

This is caused by the atomic structure of ferromagnetic bodies that enables an organized alignment of the electrons' magnetic moment in the presence of an external excitation. The material can be considered as constituted in areas forming magnetically neutral zones (Weiss magnetic domains), and when they are externally excited this equilibrium is broken by the orientation of the magnetic moments in the direction of this excitation.

Knowing the curve of initial magnetization and the hysteresis loop (Fig. 5.4) of an alloy gives information on the values of the magnetization and demagnetization fields. A wide field is the characteristic of a "hard" magnetic material, which will demagnetize with difficulty (used as permanent magnets). On the other hand, a "soft" magnetic material with a narrow hysteresis loop will be used as an electromagnet and is easily demagnetized.

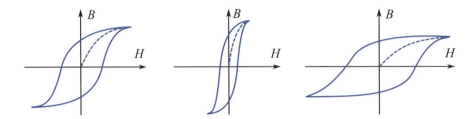

Fig. 5.4 Examples of standard, soft, and hard magnetic material hysteresis loops

Note that magnetic testing creates a remanent induction in the material. It is often necessary to proceed to demagnetization by decreasing hysteresis loops in order to return this material to a magnetic neutral state. Heating or a thermal treatment exceeding the Curie point produces the same effect.

5.1.2   Different types of magnetic materials

Nickel, cobalt, and iron are ferromagnetic. Their metallurgic composition has a direct influence on their ferromagnetic properties. In the same way, the different phases of a steel are not magnetically identical. The austenitic phase is not ferromagnetic, which makes magnetic testing impossible for austenitic stainless steel grades.

Generally, the more a steel is alloyed, the more it is necessary to reach a high excitation value to obtain saturation. Heat treatments also play a role in magnetic properties: quenching operations increase the excitation value that enables saturation. Reciprocally, an alloy that saturates with high excitation values will be difficult to demagnetize because of the presence of high remanent induction, or coercive excitation, or both at the same time.

The comparison of the 2 curves in Figure 5.5 shows that a peak is reached in the curve $\mu_r = f(H)$ for a value of H corresponding to the maximum slope zone of curve $B = f(H)$. Beyond, when saturation increases, μ_r decreases, and the curves $B = f(H)$ draw closer.

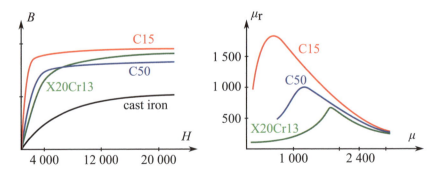

Fig. 5.5 Curves of initial magnetization of different materials

This shows that when alloys of different natures are brought to saturation, their properties tend to get closer, though they are very different when in the linear zone. Magnetic testing relies on this material property "becoming similar" when the excitation that is applied corresponds to saturation. This is one of the reasons for the robustness of the method, because excitation can thus vary without significant consequence.

5.1.3 Principle of magnetic testing

Magnetic leakage flux
The principle of magnetic testing consists in bringing an alloy to saturation by a sufficiently large magnetic field in order to create a distortion or leak in the field perpendicularly to the discontinuities that are present. This leak makes detection of the discontinuity possible.

In the absence of discontinuity, the lines of the excitation field are globally going to follow the surface of the magnetized part, and their concentration is all the higher in the part as the part's μ_r is high (the part is generally in the air, which is non-magnetic, therefore $\mu_r = 1$).

If there is a discontinuity on the surface, two configurations are possible (Fig. 5.6):
- If the material is not saturated, the field lines are distributed in the part, with a small amount (as a proportion of the whole) crossing the discontinuity.
- But if the excitation has caused saturation, a low local μ_r prevents the field lines from being channeled in the part, and a large amount (as a proportion of the whole) will cross the discontinuity and expand into the air at the surface of the material where it will be detected. This is the notion of *leakage flux*.

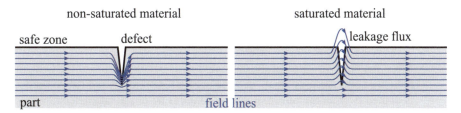

Fig. 5.6 Influence of excitation: non-saturated metal (left) and saturated (right)

This expansion of the field lines mostly located at the perpendicular of the discontinuity is going to locally modify their normal and tangential components. The presence of a discontinuity causes a local reduction of the tangential component and the apparition of a differential component of the normal excitation.

Two modes of detection can be considered:
- The measurement of variations in the normal and tangential components with sensors. This technique is applied to parts with a simple geometry, such as bars, tubes, and wires. The method is then called method by magnetic flux leakage testing.

- Discontinuity visualization, where a ferromagnetic powder is spread over the surface of the part and accumulates at the level of the leak flux, at the vertical of the discontinuity. This technique is called magnetic particle testing.

The phenomenon of magnetic particle image apparition

The normal excitation field resulting from the leakage flux at the vertical of the discontinuity creates a force of attraction for the particles of the ferromagnetic powder that was spread. More precisely, it is the gradient of the excitation field that creates this attraction force F:

$$F = 1/2 \ X \ T \ \mu \ \mathbf{grad}(H^2)$$

with X magnetic susceptibility of the particle $= -1$
         T volume of the particle
         H leak field exerting an action on the particle
         μ magnetic permeability of the particle.

The force is even higher if μ is high and/or if the defect is narrow (higher gradient). As the gradient is differential, there are actually two forces and two accumulations close to the two edges of the discontinuity. These accumulations will create a visual contrast between the material surface and the discontinuity zone. As this zone is usually "thin," both accumulations merge and the resulting image, for a linear defect, appears as a thin line.

Influence of the excitation field direction

When a field is applied to a part (Fig. 5.7), it magnetizes the part in one main direction. Two situations may then occur:
- When the main direction of the discontinuity to detect is perpendicular to that of the field, the field gradients are high, and the accumulation of the indicative product is large and appears all over the length of the discontinuity. The image is clear; it is a good quality detection.
- When the main direction of the discontinuity is parallel to that of the field, the field gradients are very small, and the accumulation of the indicative product is diluted over an infinitesimal length. There is no image and detection is not possible.

For example, on the surface separating steel ($\mu_{r1} = 500$) from air ($\mu_{r2} = 1$):
- continuity of the induction normal component $B_1 \cos \alpha_1 = B_2 \cos \alpha_2$
- continuity of the field tangential component $H_1 \sin \alpha_1 = H_2 \sin \alpha_2$

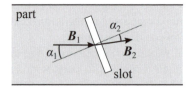

Fig. 5.7 Influence of defect orientation

Hence $H_1/B_1 \tan \alpha_1 = H_2/B_2 \tan \alpha_2$, that is, with $B = \mu_r H$ and $\tan \alpha_2 = \mu_{r2}/\mu_{r1}$ $\tan \alpha_1$, the following is deduced:

1) if B_1 is perpendicular to the crack ($\alpha_1 = 0°$):
   $\tan \alpha_1 = \tan \alpha_2 = 0$ therefore $\alpha_2 = 0°$ and $B_2 = B_1$
   therefore $\mu_{r2}H_2 = \mu_{r1}H_1 \Rightarrow H_2 = 500\ H_1$

2) if B_1 is at 45° to the crack ($\alpha_1 = 45°$):
   $\tan \alpha_1 = 1$ and $\tan \alpha_2 = 1/500 = 0.002$
   therefore $\alpha_2 = 0.11°$ and $B_2 = 0.707\ B_1$
   therefore $\mu_{r2}H_2 = 0.707\ \mu_{r1}H_1 \Rightarrow H_2 = 350\ H_1$

3) if B_1 is almost parallel to the crack ($\alpha_1 = 89°$):
   $\tan \alpha_1 = 57$ and $\tan \alpha_2 = 0.115$
   therefore $\alpha_2 = 6,5°$ and $B_2 = 0.017\ B_1$
   therefore $\mu_{r2}H_2 = 0,017\ \mu_{r1}H_1 \Rightarrow H_2 = 9\ H_1$.

As a conclusion, when the angle between discontinuity and the main direction of the excitation field is under 90°, successful detection tends to decrease, and the image is less clear. Detection is considered as acceptable until 60°.

In general, the application of the excitation field in a direction perpendicular to the supposed direction of the discontinuities is to be favored. When there is no presupposed knowledge of this direction, it is necessary to execute two tests, with excitation fields orientated at 90° which, at worst, results in the possibility of a test at 45°.

Influence of the nature of the field, penetration depth
Usually, the magnetic field used to magnetize a material or the part to inspect is a sinusoidal alternating field type, as it comes directly from mains power. However, its use generates induction phenomena and limits the penetration depth into the part: this is the *skin effect*. A magnetization of the sinusoidal alternating type will thus limit the detection of discontinuities positioned at about 1 mm, that is, almost exclusively emerging ones.

The use of a direct magnetization field in the form (Fig. 5.8) of a rectified wave (half-wave or full-wave rectification) can detect non-emerging discontinuities located a little deeper. But this method is of no real help in going further, as internal detection will not be possible for a part of a few tenths of a millimeter. There are two main reasons for this:
- The field lines of an internal defect will be distributed throughout the section and the leakage flux, unlike alternative magnetization, which will be small and diluted.
- For an internal discontinuity, reaching the same field density as that obtained with an alternating field on the surface would require field values that are too high, which could be destructive because of the temperature rise generated in the part.

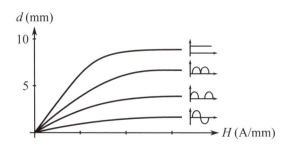

Fig. 5.8 Detection depth in a light alloy steel

Magnetic particle testing must then be considered as a surface or sub-surface test method (approximately 10 mm). The only reason to use a non-alternating field is to confirm the emerging character of a discontinuity.

Magnetization by flux application or by current application

Magnetization can be carried out in two different ways (Fig. 5.9):

- The field is directly applied in the part by a device producing this excitation (current, electromagnet). This is the technique of *flux magnetization*, also called longitudinal magnetization, for transverse defects.
- A current is run in the part, generating an excitation field that "self-magnetizes" the material to inspect. This is the *current magnetization* method, also called transverse magnetization, for longitudinal defects.

Some installations combine the simultaneous application of longitudinal and transversal fields, making it possible to visualize defects with opposite orientations.

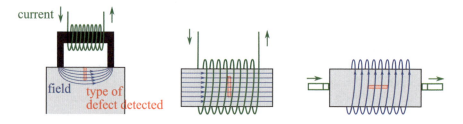

Fig. 5.9 The different modes of magnetization, by flux magnetization (with an electromagnet or a coil) and by current magnetization. The field must be perpendicular to the defect.

5.1.4 Optical phenomena and vision

In most cases, inspection is carried out visually. Detection and reading of indications are included in the visible spectrum, and the characteristics of human vision relate directly to test results. Magnetic powders are either black indicative products, visible by reflection in white light, or fluorescent products visible when excited by ultraviolet light, also called black light.

Vision

Detection of indications obtained in magnetic testing relies on contrast and luminosity phenomena associated with the way the eye works. Perception is maximized when there is contrast in luminosity (black and white) and opposition between colors. This fact sometimes requires the addition, during white light testing, of a white primer on the tested surface, as a contrast surface against the black image due to the powder accumulation by the discontinuity.

Use of white light

When a test is carried out using a black indicative product under white light, the product absorbs all the used wavelengths. If a white primer is used, or just the simple reflecting surface of the part, there will be a reflection of all the wavelengths. But the white light used to light the surface has to be strong enough so that this reflected light sensitizes the retina. Minimum values for lighting are recommended in the standards.

Use of fluorescent light

Fluorescence is a particular case of luminescence (radiation in the presence of energy input), which corresponds to a synchronized phenomenon of radiation emitted by excitation. When a test is carried out with fluorescent products, an ultraviolet light is used to excite the product in a dark environment.

This product (and only this one) radiates a green-yellow color corresponding to a wavelength of 550 nm (the maximum sensitivity of the human eye) with a strong contrast against the surrounding dark. The resolution power is increased by a factor of 2 or 3 compared to a test in white light. This emitted light has to be strong enough to sensitize the retina. Minimum values for energetic lighting are recommended in the standards.

5.2 TECHNOLOGY AND MATERIALS

5.2.1 Indicative products

There are a few dozen micron particles (made of ferromagnetic material with a high relative magnetic permeability μ_r) that can be easily demagnetized without significant remanence, such as magnetite ($Fe_2 \_ O_4$). Their color is either grey/black, visible in white light, or colored with a fluorescent product to react with ultraviolet light. Their granulometry is smaller than for fluorescent products.

Indicative products have different presentations:
- Dry powder, essentially for manual application
- In suspension in oil-based or aqueous-based liquid. In the latter case, they are associated with a wetting agent and an antioxidant.

When the products are in liquid suspension, they can be conditioned:
* In a support liquid or as paste to dilute before use
* In sprays for direct use

Granulometry and concentration are chosen depending on the surface condition of the part, the type of searched defect, the modes of application, and lighting. Classic products with support liquid are operational up to a temperature of about 80°C.

For example, when searching for cracks with a very narrow opening, it is better to use thin fluorescent powders, able with only a few grains to radiate a light that can be detected by the eye. Whereas a few grains of grey powder, even contrasting against a white paint coating, will remain difficult to detect.

5.2.2   Contrasting primer and solvent

It is possible to apply a thin coat of white background to increase contrast with indicative product. This can be found as primer in a spray can. The product dries in a few dozen seconds so that the indicative product can be applied right after. This primer can be removed with a special solvent if necessary.

5.2.3   Lighting equipment

For colored penetrants, white light lighting can be achieved in daylight, or using incandescent lamps, diode lamps, or xenon discharge lamps. Inspections should be carried out with a lighting level exceeding 500 lux.

For fluorescent penetrants, ultraviolet lighting can be achieved using mercury vapor lamps called *Wood's lamps* or light-emitting diodes (LED). Stray white light must be limited to 20 lux.

5.2.4   Reference parts

There are different types of reference or standard parts (see EN ISO 9934-2 standard), notably to check the quality of the indicative products. Some are thick steel disks, with a hole in the center and cracked by quenching (type 1 standard, Fig. 5.10). They are magnetized by a central conductor, and the product evaluation is carried out by visual comparison.

Fig. 5.10 Type 1 and 2 standard parts, ISO 9934-2 standard (SREM)

Devices constituted with two magnets in opposition magnetizing a split bar can also be found (type 2 indicator, Fig. 5.10). The length of the signature obtained is used to evaluate the product performance. There are also parts with a small diameter through holes of different depths; these have to be magnetized by an external source but the results are poorly representative.

5.2.5 Magnetization indicators

Other parts, called magnetization or field indicators (AFNOR, Berthold, ASME), are designed to be placed on the examined part. These indicators are made of ferromagnetic material and include artificial emerging defects. They are used to check the capability and reliability of the method, but they do not measure the excitation field on the surface, which is the only way to check the real magnetization condition of the part. The ASME indicator, for example, is made with 8 triangles brazed together side by side. It is covered by a copper film. In the same way as for the BERTHOLD cross, the most visible lines are those correctly orientated, perpendicular to the force lines (Fig. 5.11).

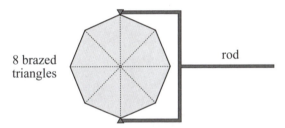

Fig. 5.11 ASME magnetization indicator

5.2.6 Tangential field gauges

Saturation "indicators" do not exist. Only a device measuring the tangential excitation field at the surface of the part (Fig. 5.12), when magnetizing, will make it possible to quantitatively check that the material is in a correct condition of magnetic saturation. This is achieved by a semi-conductor sensor to be positioned on the surface of the

Fig. 5.12 Hall effect sensors (SOFRANEL)

part, connected by a cable to the device that gives the value and/or the shape of the tangential field.

5.2.7   Equipment: magnetic testing devices

Except for permanent magnets, meters work with electrical supply, generally from mains power. They can be mobile or installed as a fixed bench. They are electrically insulated from the mains and have an integrated transformer.

Current magnetization devices
These generators supply a high current with a low voltage. The current is applied to the part through electrodes (Fig. 5.13). There are several available types of current: rectified with two alternations (often called "direct"), rectified with one alternation (or "pulsed"), and sinusoidal alternating current. The demagnetization function is achieved by a decreasing alternating current (for 10 seconds approx.)

Fig. 5.13 Magnetization by mobile electrodes (SOFRANEL)

Permanent magnets and electromagnets
A permanent magnet tester (Fig. 5.14a) is adapted for difficult to access zones or those with a risk of fire and/or explosion. This is the case when it is hazardous to use electrical currents, as in the inspection of oil tanker tanks. The sheet metal plates must not be too thick, less than 10 mm approx.

Magnets and electromagnets are able to directly transmit the magnetic field or flux to the part to be tested. They are constituted with U-shaped armatures and are often articulated (Fig. 5.14b).

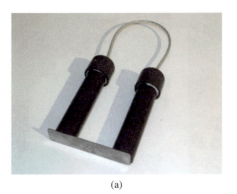

(a)   (b)

Fig. 5.14 Magnetization by permanent magnets (a), with a plate to close the circuit and by portable electromagnet (b), with integrated UV lamp (SOFRANEL)

Cables and coils

These are cables or coils (Fig. 5.15) energized by a current generator, which creates a magnetic field in their close surroundings. They are used to magnetize the nearby parts or inside for a winding.

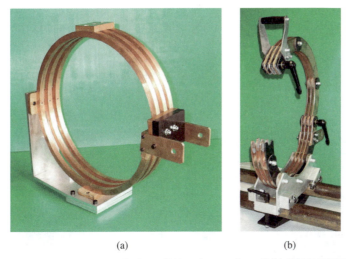

(a)   (b)

Fig. 5.15 One-piece magnetization coil (a) and an opening coil (b) (SOFRANEL)

Testing benches

These are fixed devices used to carry out current or flux magnetization in a part. They are equipped with a high energy generator, and it is possible to test parts with a large length and section.

The part is held by electro-pneumatic actuators, and loading, unloading, and testing can be automated giving the possibility of integration into a manufacturing line (Fig. 5.16).

(a)   (b)

Fig. 5.16 Testing benches for crankshaft (a) and spraying system (b) (SREM)

Demagnetization tunnels

These are coils that can supply a large alternating excitation field. They are used manually or inserted into an automated process after a testing bench (Fig. 5.17). When a part is inserted into the coil, it is subjected to a large magnetic field, which then strongly decreases as the part moves away, the purpose being to demagnetize the part.

Fig. 5.17 Demagnetization tunnels (SREM)

5.3   TESTING METHODS

5.3.1   Test operations

Magnetic testing is carried out in the following chronological way:
- Check that the part is magnetic (reference document, use of a magnet)
- Surface preparation (cleaning, application of a contrasting primer)
- Check the performance of the indicative product and lighting
- Simultaneous magnetization and check of the obtained tangential magnetic field
- Application of the indicative product
- Stop magnetization, then observation, interpretation

5.3.2   Choice of the type of magnetization: flux or current

The choice of the type of magnetization depends on several criteria, the first being that the main direction of the field is perpendicular to the supposed direction of the defect. There are other parameters to take into account, too:
- Type of defect and metallurgic characteristics of the part: a deeper penetration is reached when using the current magnetization technique.
- Shape of the part and dimensions that may prevent the use of a coil, for example. Note that there is a technique that consists in temporarily "extending" the part (Fig. 5.18) to obtain a better orientation of the field lines compared to possible defects.
- Surface condition, quality of contacts: current magnetization is to be avoided in order to prevent damaging parts with a very good surface condition.

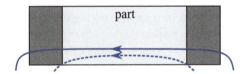

Fig. 5.18 Addition (during testing) of two external masses

The applied tangential field will depend on the metallurgical characteristics (initial magnetization curve), a highly alloyed steel requiring a higher excitation field value than a light alloy steel. Recommended minimum values can be found in the AFNOR standards:
- The applied tangential field must exceed or be equal to 2000 A/m
- The induction reached by the material must be of 1 tesla

5.3.3   Choice of the test conditions

The inspection can be carried out by applying the indicative product and the magnetization at the same time, taking care to stop spraying before stopping the field in order to avoid "erasing" the signature: this is called the simultaneous method. Generally the magnetization time does not exceed a few seconds. Several sequences of magnetization can be carried out successively in order to avoid heating the part (particularly on the passage of a current).

The inspection can also be carried out by application of the product after the magnetization: this is known as the remanence method.

5.3.4   Hygiene, safety, and environment

The use of indicative products and liquid carriers can be damaging to the environment and health. They require specific conditions of use, packaging, storage, and recycling.

The ultraviolet radiation used to excite fluorescent penetrants belongs to the UV-A range (320 nm to 400 nm) and has a limited effect on the skin at the levels of the energetic lighting used. However, it is advisable for operators to avoid exposing their skin over long periods.

As the magnetic fields used are significant, operators must be exposed to these fields as little as possible. When people are exposed to low frequency magnetic fields (for example, 50 Hz), the limit value of exposure (in current density) through the human body must not exceed fixed values (European directive), which for the concerned frequency range is at 10 mA/m^2, a value difficult to measure. This is why safety agencies like the ICNIRP recommend a 1-tesla limit value for induction fields.

As magnetic testing uses high currents, electrical safety is ensured by the electrical insulation between the current used in the part and main supply. The equipment must be compliant with regulations and must be periodically checked.

5.3.5   Interpretation and acceptance criteria

The apparition of an indication in magnetic testing being almost synchronous with magnetization, it is advisable to inspect the surface of the part at the beginning of this phase. The inspection must allow you to distinguish between false indications and indications actually caused by a defect.

A diffuse indication in the proximity of a modification in the part geometry (fillet, embossment, etc.) can be the sign of a mechanical accumulation of powder, or excessive saturation. Local variations of magnetic permeability, caused by zones with different metallurgic properties, can also be a source of false indications. A counter-test with a lower excitation and/or a better draining of the excessive indicative product in liquid phase can help reach a better "signal to noise ratio." In the same manner, an excessively high excitation field on forged parts can reveal the fiber drawing.

The only reliable indication is the length of the indication of a surface or sub-surface defect. For the rest, experience (with caution) can be useful:

- A wide indication is generally the sign of an underlying indication and/or with a character that is not only linear
- A neat thin indication is the sign of an open defect, such as a crack

As in penetrant testing, acceptance criteria depend on the products, codes, or recommendations specific to a sector of activity with some common points in their approach:

- Indications are considered linear if they are 3 times longer than they are wide. They can be independent or aligned.
- Otherwise, indications are non-linear and can be aligned or grouped according to whether a line or a mass is formed.

The acceptance of these indications will depend on their dimensions, their quantity over a given surface (accumulation), their position, the thickness of the part, and the combination of these various characteristics.

5.4   EXAMPLES OF APPLICATIONS

5.4.1   Inspection of molded parts

The defects detectable at this stage (Fig. 5.19) are mainly tears, retraction cracks, and "shrinkage cavities" flush with the surface. More occasional defects can also be detected, such as blowholes, pitting, and cold shuts.

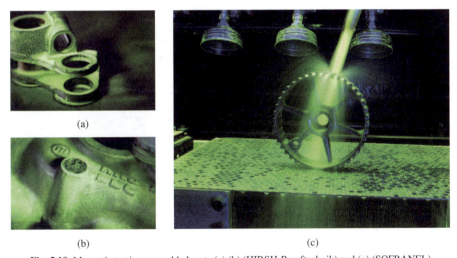

(a)

(b)   (c)

Fig. 5.19 Magnetic testing on molded parts (a) (b) (HIRSH-Prueftechnik) and (c) (SOFRANEL)

5.4.2   Inspection of forged and laminated parts

Magnetic testing is particularly suitable for testing products at this stage of the elaboration of a product and after the lamination operation. Linear defects can be seen, caused by the "upwelling to the surface" of internal defects such as inclusions or porosities, folds, tears, cracks, etc. For forged products (Fig. 5.20), forging lines and folds, tears, quenching, and shrinkage cracks are detectable.

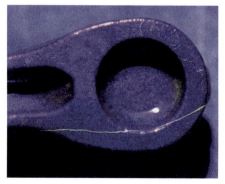

Fig. 5.20 Magnetic testing on forged parts: pivot (SOFRANEL) and connecting rod (SREM)

5.4.3 Inspection of machined and treated parts

The main defects that can be detected after machining are scratches and grinding cracks, and after a thermal treatment (Fig. 5.21), quenching and shrinkage cracks.

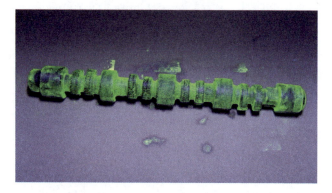

Fig. 5.21 Crack detection on a camshaft (SOFRANEL)

5.4.4 Inspection of welded joints

Detectable defects likely to concern welded joints: transversal and longitudinal cracks at the surface of the joint or the HAZ, emerging porosities, bad junctions, or cracks at the interface of the weld and base metal. Defects can also be detected on sheet metal plates (splitting) before welding. Lack of penetration can also be detected on joints with a small thickness, and possibly, weld undercuts that are normally detected by visual testing.

Magnetic testing is a method widely used for weld testing (Fig. 5.22).

Fig. 5.22 Magnetic testing on different welds (CNDR)

5.4.5   Inspection of parts in service

The main defects detectable after use are fatigue cracks, tensile cracks, torsion cracks, shock cracks, as well as cracks caused by over-rapid heating or cooling. This type of defect is quite suitable for magnetic testing detection (Fig. 5.23).

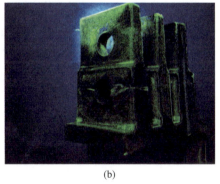

(a)   (b)

Fig. 5.23 Fatigue cracks on a nut (a) (INSACAST) and on a yoke (b) (CNDR)

5.4.6   Special applications

Magnetic testing can be used in sub-aquatic testing. In this case, an electromagnet is used, supplied by a long cable linking the magnetization device to the generator, which remains out of the water. In this technique, ultraviolet lighting or white light is used, and in this case, the indicative product is red.

Magnetic testing can also be used in high temperature conditions with dry indicative products, up to 200 or 300°C. For low temperatures, tests can be carried out down to –40°C with oil-based products.

5.5   CURRENT AND FUTURE POTENTIAL

In theory, magnetic testing is in competition with penetrant testing for ferromagnetic parts. Their detection sensitivity is equivalent. Although penetrant testing is to be favored for parts with complex shapes, magnetic testing is profitable in cases of bad surface conditions: the use of alternating currents transmits vibrations to the powders and makes them "jump over" the obstacles.

Magnetic testing is also faster to implement (a possibility of two to three minutes, as opposed to a minimum of one hour for any penetrant testing operation). Recycling is also much easier than with penetrant testing, as the powders are magnetic.

Magnetic testing is still being improved and there are several possible orientations.

5.5.1   Quality, performance, and impact on the environment

Product performance is the subject of constant improvement, in brightness, sensitivity, and reliability. Manufacturers are adapting the characteristics of the products and associated equipment to the regulations, particularly in the fields of environment and safety, in the propellants of spray cans, for example.

5.5.2   Associated equipment

Progress has been made and improvement continues in current generators: the functions of intensity adjustment and control have been automated and improved in quality thanks to progress in electronic device integration. The improvement of electro-technical sub-assemblies has made a reduction in weight of the different magnetization pieces of equipment possible. Inspection benches have also benefited from progress; their size has been reduced, they are available in modular designs, and they require less energy.

Excitation field testers have evolved, too. The trend is toward the integration of functionalities that can measure true efficient value whatever the wave form and increasingly offering a visualization of the signal.

5.5.3   Excitation and observation of indications

Fluorescent penetrants were initially designed to be excited by an ultraviolet light of 365 nm. Using an excitation light in a spectral range of 405 nm to 450 nm (visible field: violet/blue) would help solve the problem of the risks caused by UV-As on eyes and epidermis.

5.5.4   Rotational fields without contact

With these systems, a part can be subjected to rotating magnetic fields in a magnetization chamber. Omnidirectional magnetization ensures the detection of all the defects detectable in magnetic testing. It is likely to become a future device for automatic testing.

5.5.5 Magnetization frequencies

Today's devices generate magnetic fields using mains frequency (50 to 60 Hz). There are studies in progress in aiming to identify performance in the use of harmonic frequencies of several hundreds of Hz, to increase detectability while reducing the level of necessary energy.

5.5.6 Automatic detection

The goal in the improvement of this method is not new, as devices able to identify indications and distinguish between images of discontinuities and false indications, with digital recording and adequate processing, remain a matter for research. These devices mainly concern inspection using fluorescent products. While background interference problems are less important than for penetrant testing, parts with complex shapes or false luminous indications remain obstacles for an industrial use of this approach.

5.5.7 Crack depth

Some research has tried to link the measurements of the tangential field and particularly the passage to zero (x_0) at the depth of the defect, but no real industrial success has followed yet (Fig. 5.24). It is an interesting idea, but there has been no validation, especially due to the difficulty in creating similar real cracks for which only the depth characteristic is different.

Some developments in magnetic testing techniques are very similar to other electromagnetic methods, among which the ACFM method. Chapter 12, entitled "Other NDT methods," provides an overview of these methods.

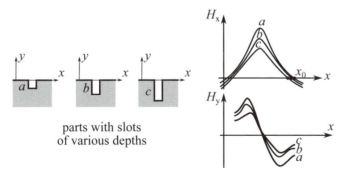

Fig. 5.24 Influence of crack depth on the measured tangential field

5.6   REFERENCES

CHERFAOUI M., Essais non destructifs, *Techniques de l'Ingénieur*, Ref. BM6450 V1, July 2006.

COFREND, *Magnétoscopie: annales officielles de la certification, niveaux 1, 2 et 3*, Editions Lexitis, 2014.

COFREND, *Guide des bonnes pratiques d'éclairage en ressuage et magnétoscopie*, Les Cahiers Techniques de la Cofrend, Editions Lexitis, 2014.

DUBOSC P., CHEMIN P., *Magnétoscopie, Manuel de cours niveaux 1, 2 et 3*, Les Cahiers Techniques de la Cofrend, Editions Lexitis, 2013.

GRAVELEAU S., CHEMIN P., *Magnétoscopie: aspects théoriques et réglementaires, Techniques de l'Ingénieur*, R6202, 2015.

INSAVALOR, *Contrôle Non Destructif par Magnétoscopie niveaux 2 et 3*, Lecture documents, 2014.

SCHMIDT J.T., *Non-Destructive Testing Handbook, Volume 6, Magnetic Particle Testing*, ASNT, 1989.

TOITOT M., *Magnétisme avec ses applications à la magnétoscopie*, Les Cahiers Techniques de la Cofrend, Editions Lexitis, 2013.

NF EN ISO 9934-1, *Magnétoscopie, Partie 1: Principes généraux*, October 2015.

NF EN ISO 9934-2, *Magnétoscopie, Partie 2: Produits magnétoscopiques*, October 2015.

NF EN ISO 3059, *Contrôle par ressuage et contrôle par magnétoscopie, Conditions d'observation*, February 2013.

CHAPTER 6

RADIOGRAPHIC TESTING

A flux of photons (X or γ) penetrates a material and is attenuated by this interaction. It leaves some energy in the detector (blackening analog film, for example) and causes a specific gray level in the image. The presence of a defect, when it generates a difference in attenuation compared to sound matter, will show as a different level of gray (Fig. 6.1). In radiographic testing, you must have access to the two sides of the part. The principle is the same whether the detection support is analog film or a digital system (digital radiography, radioscopy).

There are three types of interaction with matter, and the different respective probabilities are predictable according to the characteristics of the incident beam and the material. These interactions help predict, via a global attenuation law, the safety conditions and time of exposure according to the required image quality. The latter is the ability of an image to help detect a possible defect and depends on parameters related to detectors, the desired (or imposed) contrast, and to geometric blurring. This blur is, among other things, related to the size of the photon source.

X-ray (XR) tests are preferred to gamma-ray tests (γR) for the quality of the image obtained (smaller source, even micro-focus of a few μm), but the portability of the γ devices is more appreciated when working on-site.

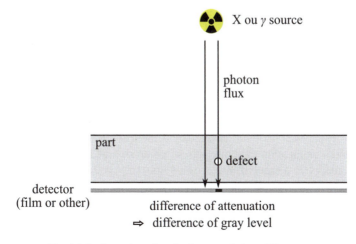

Fig. 6.1 Radiography or imaging by transmission of X or γ rays

The image from X-ray testing can be archived and is relatively easy to interpret. But it does not help, in principle, for an easy location of the defect in depth and only detects plane defects (cracks) when they are correctly oriented (parallel) with the photon flux.

This inspection process is applicable to all materials, to all manufacturing processes, and throughout the whole lifetime of the part. It can be difficult to implement on thick parts. It is dangerous (photon – human body interactions), and it requires qualified personnel (safety, zoning, etc.).

6.1   PHYSICAL PRINCIPLES

6.1.1   Ionizing radiation

Particle or electromagnetic *radiation* is *ionizing* when it is likely to remove electrons from matter. To do so, the radiated energy has to be higher than the electron binding energy (it is considered as ionizing when it exceeds 10 keV approx.). The most frequent forms of radiation are alpha, beta, and neutron rays, as well as photons (γ and X).

X and gamma (γ) rays are electromagnetic forms of radiation, just like rays of light, microwaves, or Hertz waves (Fig. 6.2). The difference comes from the way they are generated. They are not mass particle radiations (α, β, neutrons).

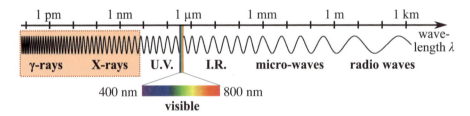

Fig. 6.2 (Arbitrary) distribution of electromagnetic radiation

The wave aspect helps in defining a wavelength (a very small one here, much lower than 10^{-9} m) and handling the characteristics of propagation in a vacuum. The corpuscular aspect (in terms of quanta or flux of energy) is more commonly used to define the attenuation laws. So, electromagnetic radiation can be represented as a photon flux where each photon is charged with energy ($E = hc/\lambda$ with h being Planck's constant) and a momentum (a photon at rest has no significance).

Their energy is measured in electron-volts (1 eV = 1.6×10^{-19} J). X-rays are emitted from electronic interactions and γ rays from nuclear interactions. X-rays are thus generally polychromatic (or poly-energetic) because of the way they are produced.

The propagation of a photon in a vacuum represents a straight line without interaction (hence the word "ray"). A beam is then an assembly of rays. The power radiated by a source is measured by energetic flux (in eV·s^{-1}).

6.1.2 Interaction of photons with matter

The representation of photons is only justified when there is an exchange with matter. In general, photons have little interaction with matter, and a few of them go through matter without even interacting (directly transmitted radiation). *Interactions* can occur according to three main mechanisms (the Compton, photoelectric, and pair effects), though in practice, there are two major types in the energy range used in NDT, that is, energies comprised between 1 keV and 1 MeV: the photoelectric effect and Compton scattering.

These interactions can be predicted when the incident beam and crossed material are known. This will help forecast a global attenuation law, which will be used to choose the setting parameters for the radiographic test.

Compton effect (Compton scattering)

The *Compton effect* is an interaction with a transfer of energy: part of the energy from the incident photon is transferred to an electron, generally one with a loose binding, which is ejected from the atom (Fig. 6.3). The energy of the scattered photon E_d is thus strictly lower than that of the incident photon E_0 and depends on the angle θ under which the scattered photon is emitted.

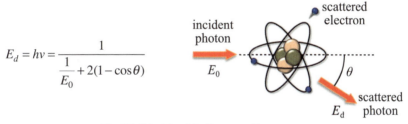

$$E_d = h\nu = \frac{1}{\dfrac{1}{E_0} + 2(1 - \cos\theta)}$$

Fig. 6.3 Principle of the Compton effect

The energy of the scattered photon is maximal when $\theta \to 0$ and minimal when $\theta = \pi$ (these results are used when designing the thickness of bunker walls). When the diffusion angle θ is higher than $\pi/2$, the photon is then called a *backscattered photon*. For $\theta = \pi/2$, if $E_0 = 10$ MeV, $E_d \approx 5\%$ E_0 is found, but if $E_0 = 0.25$ MeV, $E_d \approx 75\%$ E_0.

Scatter perturbs the image (a noise appears) because some photons take part in the blackening of the detecting support without having followed a direct path, that is, without bringing any data about a possible defect (Fig. 6.4). As a first step, it can be deduced that to suppress scattering and therefore obtain better quality images, it is more suitable either to work with high E_0 values or in a domain where the Compton effect is not major (at low energy E_0).

Photoelectric effect

The *photoelectric effect* is a phenomenon of total absorption of the incident X photon.

The energy of the X photon is integrally transferred to an atom, which ejects an electron, generally a tightly bound one (Fig. 6.5). Atomic recoil follows and the atom is now ionized.

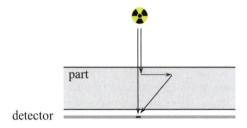

Fig. 6.4 Scatter blackens the film without producing any latent image

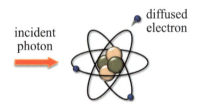

Fig. 6.5 Principle of photoelectric effect

At each step in this successive rearrangement of the electron configuration, an atom releases energy, corresponding to the difference in the binding energy of the two concerned electronic layers, either in the form of radiation called *fluorescent radiation*, or in an electronic form by ejecting an outer electron; this is called the Auger effect.

The output of fluorescence increases with the atomic number of the material. The photoelectric effect is most prevalent when energy is low, typically up to a hundred of keV (this limit depends on the material and increases with the atomic number).

Pair production effect

The *pair production effect* refers to an X photon penetrating the electrical field in the atom and simultaneously generating 2 elementary matter particles: a positron and an electron (Fig. 6.6).

According to the law of conservation of energy, the incident energy must be at least two times higher than the mass of the idle electron (= positron), that is:

$$E_0 \geq 2\, m_0 c^2 = 2 \times 0.511 = 1.02 \text{ MeV.}$$

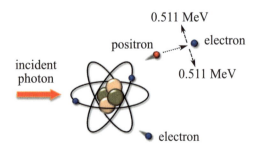

Fig. 6.6 Principle of pair production effect

Distribution is instead made in the forward direction. After slowing in the matter and loss of kinetic energy, the positron recombines with an electron to give two X photons of 0.511 MeV in two opposite directions. This interaction is impossible with the element ^{192}Ir (§ 6.2.1), which is too weak and not likely to be encountered in the field of energy used in standard NDT.

6.1.3 Attenuation laws

When a beam of photons impinges a material with a normal incidence, a part of it is going to interact. A probability of interaction μ can be defined per unit of length. This coefficient μ called ***linear attenuation coefficient*** is thus linear with density ρ of the material. It is also the sum of the probabilities of each interaction per unit of length (Fig. 6.7), that is:

$$\mu \approx \mu_{\text{Photoelectric}} + \mu_{\text{Compton}} \; (+ \text{ possible } \mu_{\text{Pair}})$$

Multiple measurements of these coefficients have been made and are now known for all elements and all photon energies. The inverse of the linear attenuation coefficient $\lambda = 1/\mu$ is the mean free path: it represents the average distance travelled by a photon before interacting with matter. The ratio μ/ρ (mass attenuation coefficient) is often used as it is independent from ρ and it proves to be constant too, for a given material, independently from its state (solid, liquid, or gaseous).

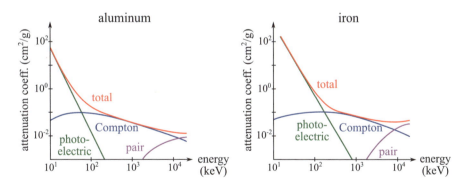

Fig. 6.7 Attenuation coefficients μ/ρ, according to energy E, for aluminum and iron

In relation with the energy of the incident photons, one interaction is clearly dominant compared to the others. Figure 6.8 shows these areas, depending on energy and the chemical composition of the material.

If the material is homogeneous and the radiation mono-energetic, the integration of interactions on a material thickness x gives the ***attenuation law of X-rays***, called the ***Beer-Lambert law***. This law defines the intensity of radiation directly transmitted, that is, the N photons (out of the N_0 incident ones) subjected to no interaction and

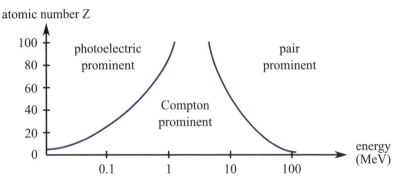

Fig. 6.8 Modes of attenuation according to the atomic number

essentially contributing to the deposit of energy on the detecting support (blackening the film, for example):

$$N = N_0 e^{-\mu x}.$$

It is also possible to consider that the energy lost by attenuation and expressed in terms of photoelectric and Compton interactions can also be considered in terms of absorption and scattering. In this case, we are directly in phase with the calculations necessary to blacken the film (time of exposure), ensure radioprotection (delimitation of zone), and guarantee a good quality image (blur notion). So, if we define dN the number of interactions, we can write:

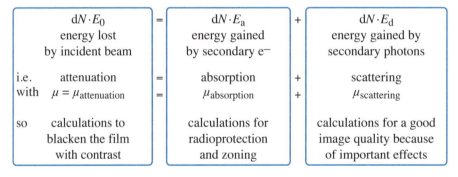

6.2   TECHNOLOGY AND EQUIPMENT

6.2.1   The production of ionizing rays

Alpha, beta, and neutron radiation, and photons rays (γ and X) have different origins, but all share the property of ionizing the matter they pass through. In general, a radiography executed with X-rays is preferable because the image quality is better than that obtained with gammagraphy. We use γ-rays when for technical reasons such as accessibility it is not possible to use X-rays.

X-ray generation

X-ray generators are electrical devices with an integrated high voltage transformer. The electrons in circulation in the filament are extracted by the very high electrical field, are accelerated, and hit the positive electrode, generally made of tungsten. Figure 6.9 shows the principle in the case of a reflective anode, that is, when the useful beam of X-rays is orthogonal to the axis of the electron beam. The sizes of the electron beam, the impact zone (thermal focal spot), and the X-ray source viewed from the detector (effective focal spot) are different and depend on the anode angle.

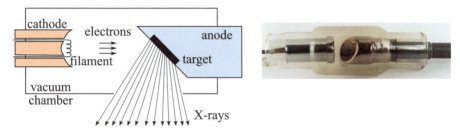

Fig. 6.9 Diagram and picture of an X-ray generator with reflective anode (COMET AG)

The kinetic energy acquired by the electrons before hitting the anode is only the result of the product of their charge by the difference in potential: for a tube working with a high constant voltage of 100 kV, all the electrons have a kinetic energy of 100 keV.

The range of X-ray generators is very wide (Fig. 6.10a); the effective focal spot can be a few millimeters in size, or less than a micron, the electrical power can be of several kilowatts or approximately of a watt (available power and focal spot sizes are often closely linked). The anode is often placed in reflection and can be panoramic. Bombarding a tungsten deposit on a carbon blade can give a photon direction similar to a "transmission," which makes higher magnifications possible (Fig. 6.10b).

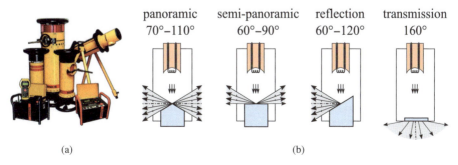

Fig. 6.10 Different X-ray generators (a) (BALTEAU) and types of anode (b)

The spectral energy of the X beam (Fig. 6.11) shows 2 components, a continuous one from the electron-nucleus interactions (***braking radiation*** or ***bremsstrahlung***) and a discrete one caused by ionization and forming a line spectrum with spectral lines that are characteristic of the anode (***fluorescent radiation***).

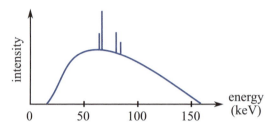

Fig. 6.11 Illustration of obtained spectra

The higher the voltage applied to the tube terminals, the faster the velocity of the electrons, and the higher the X-ray energy. The kilovolt (kV) setting represents the quality (or energy) of radiation while the milliamp setting (mA) is linked to the quantity of radiation.

Manufacturers design their own generators complying with the laws of the countries in which they are to be used. To limit the surface of the beam coming from the radiating unit, it is possible to install devices that reduce the used X-ray beam surface at the output window. These devices can be made simply by the addition of prefabricated plates or more complex mechanized systems. This set of devices is called ***collimators***.

There are other, more marginal, ways to produce X-rays, such as synchrotron radiation sources, linear accelerators, or electronic microscopes.

γ ray generation: radio-elements

Radioactivity is a physical phenomenon in which radioisotopes return to a stable state. An unstable nucleus releases its excess energy by emitting radiation. The phenomenon of energy release causes a modification of the nucleus called disintegration; this is radioactivity.

When disintegrating, a nucleus can release different types of radiation: α, β, neutrons, and photons. A form of radioactivity corresponds to each of these types of radiation. Uranium 238 is unstable, as are ***cobalt 60***, and ***iridium 192*** used in gammagraphy, or carbon 14 and potassium 40 which are present in the body.

In nature, unstable atoms gradually lose their excitation in time, whereas artificial radioactivity is generated by humans who are able to produce unstable atoms from stable elements through the use of nuclear reactions.

The ***activity*** (A) of a radioactive source refers to the number of radioactive atoms that are subjected to a nuclear transformation (disintegration) per time unit. The legal unit is the Becquerel (Bq), which corresponds to one nuclear transformation per second. The old unit (Curie (Ci), with 1 Ci = 37 GBq and 1 TBq = 27 Ci) is still used. In industrial gammagraphy, used sources have a few TBq (a few dozen Ci) activities.

The *radioactive period T* is the time it takes for the number of radioactive nuclei to be reduced by half (= initial activity/2). This period varies a lot according to the radioisotope in question. It can last less than 1 second or more than a billion years. Here are some periods of elements that are used in industrial gammagraphy, the choice depends on the period (that rules the reloading of industrials devices), on available energies (that is, the thickness of the matter to be passed through), and whether monochromatic or polychromatic emission spectrum is desired:

iridium 192:	74 days	selenium 75:	120 days	thulium 170:	127 days
cobalt 60:	5.3 years	cesium 137:	30 years	ytterbium 169:	32 days

The quality of radiation emitted during the disintegration (types of particles, energy, number, etc.) depends on the radioelement. The emitted photon spectrum is discrete and corresponds to the levels of nuclear energies of the radioelement. *Sources* are typically millimetric and cylindrical and fitted in *source-holders* (Fig. 6.12), which are integrated to "projectors," the gammagraphy equipment enabling safety and photography functions.

Fig. 6.12 Source containing the radioactive element (size 2–3 mm) (IUT GMP NIMES)

Because of its mobile nature, the GAM (a gammagraphy device using iridium and selenium) is often used on external working sites (Fig. 6.13), contrary to X-ray tubes that are more voluminous and need a source of electricity. There are approximately 800 gammagraphy devices of this kind in France. There are two types:
- GAM 80 which contains a source of 80 Curies maximum, i.e., 2.96 TBq,
- GAM 120 which contains a source of 120 Curies maximum, i.e., 4.44 TBq.

In its "storage" position, the source-holder is contained in a protective container made of depleted uranium. The depleted uranium radiates in a very low quantity, but its high density (density 19) makes it a good shield for the sources present in the GAM.

A remote control with a standard length of 10 m brings out the source in a tube to the front. By keeping the operator at a safe distance, it ensures efficient protection during the operation. In the same way, after the inspection, the remote enables the source to return into the container. A key lock system helps the device to be safely moved and stored.

Fig. 6.13 Cutaway image of educational GAM (IUT-GMP NIMES)

In France, gammagraphy devices in a company can be used only if they have been authorized by the French Nuclear Safety Authority (Autorité de Sûreté Nucléaire ASN) to possess and/or use this type of device. This authorization is valid for 5 years.

6.2.2   Image formation and detectors

Radiographic film

This is a special photographic film with a high thickness of sensitive emulsion and a high content of silver halide. The higher the radiation energy, the more the metallic salts contained in the photographic emulsion oxide, and the more the film blackens.

Films are usually stored in a paper sleeve, protected from the light or in a light-alloy strongbox (X-ray transparent), with intensifying or lead screens, or in a vacuum plastic packaging with lead screens. Photon interactions are rare, but they can create highly active electrons that blacken the film.

Intensifying screens are composed of a metallic material (usually a cardboard sheet covered with a thin layer of lead) that slightly attenuates sharpness but considerably shortens the necessary exposure time (by approx. 24 times).

Radiographic film is still present in the industrial field even if digital detectors are becoming more popular. The advantages of film are significant in terms of small pixel sizes (better image quality). Its use benefits from the delayed standardization of the digital support and sometimes artificially maintains its use: for example, the protocols for films are transposable without any precautions to the rather "flexible" photostimulable screens, as opposed to the "stiff" digital detectors that are more difficult.

The use of film is again justified by another aspect of NDT, which is the testing of individual parts or of large surfaces requiring very high precision. This is a request expressed by museums for the inspection of paintings, for example. The use of film does not limit the size of the inspected area, it suffices to position a large group side by side. The resolution is the same as that of the film, which is better than dozens of microns.

The current durability of radiographic film is still satisfactory even though its industrial production is sharply decreasing because of the digital evolution of photography. The ISO 18917 standard defines the terms of residual thiosulfate and other chemical products in treated photographic products, whose lifetime is approximately 500 years. In practice, conservation potentials of 5, 10, or 50 years are often necessary.

The use of film is also greatly linked to activity on working sites, which is itself strongly linked to the use of gammagraphy.

Photostimulable screens

A *photostimulable screen* is stand-alone, flexible, or stiff (if put in a container) and has similar characteristics to film (dimension, sensibility, and resolution), though with a much higher dynamic.

It is handled like a film, is easily digitized, and can be read again. It has low sensitivity to light and after a reset, it can be used several hundred times again. It does not need chemicals for development or conservation as opposed to film.

Inspection protocols would not have to be entirely reviewed if photostimulable screens were chosen over radiography with film. This still represents an interesting intermediary step between film and digital detectors.

The photostimulable screen is primarily used in medical radiography (*radiology*).

Digital detectors

There are two main kinds of *digital detectors*: the linear detector array (one or several lines, synchronized or not) and the matrix sensor. Although the matrix sensor obtains an image directly, the linear detector array has to be associated with a movement (translational, rotation, or both).

X-rays are not necessarily and directly converted into an electric signal, and direct conversion detectors are not predominant. Indirect conversion with optical photons is generally used: the conversion is carried out by the scintillator. Optical photons are read by standard devices used in visible imaging (CCD, photomultiplier).

The coupling of the scintillator with the optical device is made by a simple contact or via a beam of optical fibers or a set of lenses and mirrors (camera coupled with scintillator screen). The choice of scintillator depends on the application, whether luminous, spatial, or temporal efficacy, or energy resolution is favored.

Matrix detectors have at best a few dozen million pixels. The size of the pixel will determine the size of the detector: the side of the pixel can be smaller than a micron (synchroton facility cameras) or of hundreds of microns, for example, for standard flat screens. Nowadays several tens of images per second can be generated, which makes real-time 2D and even 3D inspection possible. The digitization dynamic is usually in 16 bits and can reach 20 bits.

The development of spectrometric imaging is becoming a trend. Instead of integrating the total energy deposited in a pixel followed by a scalar conversion, the boom in rapid electronics enables the individual identification of photons and by coupling to a spectrometric channel, gives vector information for each pixel. Filtering is now possible at the acquisition step.

6.2.3   Influential parameters on image quality (IQ)

Geometric blurring

X and γ rays sources are not punctual, and the impulse response of the detector is not perfect: thus, the obtained image is intrinsically blurred to a certain degree, which is caused by the combination of *geometric blurring* due to the source and internal blurring due to the detector. The geometric blurring fg (Fig. 6.14) is directly linked to the size of the source, the distance from the detector, and the thickness of the inspected part. It is represented by the following relation: $fg = d \cdot e / (D - e)$.

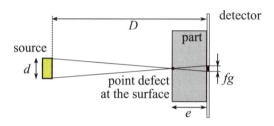

Fig. 6.14 Definition of fg, for a defect in the most critical position

In order to obtain a low amount of blurring, the source can be moved away (but exposure time will be longer). A thin part will generate a lower amount of blur, and this is the same for a defect located closer to the film. In this latter case, the study of the defect's pre-supposed location has an obvious impact on the quality of the inspection, as it makes a shot at a lower defect-film distance possible.

The dimension d depends on the source, which for X-rays are typically 0.4×0.4 mm and for γ rays 2×3 mm. This shows that by generating a lower blur, the image quality will always be better with X-rays (XR) than gamma rays (γR).

Internal blurring is directly linked to the impulse response of the detector. These two blurrings are composed by convolution, therefore the total blur is mathematically closer to the maximum value between internal blurring and geometric blurring.

Magnification

The image is obviously a magnified projection of an object as shown on Figure 6.15. The *magnification* ratio is $G = D_{SD} / D_{SP} > 1$ where D_{SD} refers to the source-detector distance and D_{SP} is the source-part distance.

In order to work with a strong magnification G, the source has to be small. Geometric blurring must be limited (for example, to the dimension of one pixel) so that it does not overtake internal blurring. This means that the source must be chosen at a size that is $(G - 1)$ times smaller than the size of a pixel. This is why using *X-ray micro-focus* is of such interest.

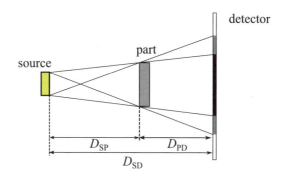

Fig. 6.15 Magnification principle in X-ray imaging and generated blurring

Density and contrast

Radiation blackens radiographic films. According to the level of attenuation, this blackening will be more or less strong and will be distributed over a scale of grey levels. Its value for a film is obtained by densitometry digitization that measures the ratio *Fi/Ft* of the incident light flux (*Fi*) on the transmitted flux (*Ft*). The (optical) **density** of the film is the decimal logarithm of the ratio *Fi/Ft*. A film of density 3 only allows 1/1000th of the light to filter through.

The **sensitometric curve** of a radiographic film refers to the curve of density according to the energy absorbed by the film or any proportional quantity, such as the absorbed dose. So, we can understand that for the same difference in received doses (AB), the contrast is better (difference of density, or blackening) in the field of high densities (Fig. 6.16).

This explains why industrial films are darker (high density is used for searching for small defects) and radiology films are lighter (for a patient, the use of low doses is compulsory). Consequently, in radiology, the use of other types of detectors is preferred in order to detect small defects.

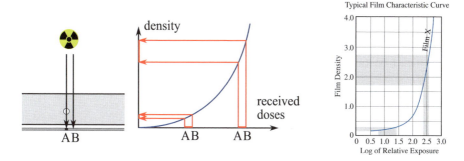

Fig. 6.16 Optimal density – principle (on the left) and real case (on the right)

In both cases, but especially in the industrial field, the use of a negatoscope (Fig. 6.17) is necessary. Three characteristics have to be taken into account: the color temperature of the tubes, the homogeneity of the light surface, and luminance. A densitometer can be used to measure the real density of films and help in the interpretation of images.

Fig. 6.17 Negatoscopes (SOFRANEL)

Scatter

Scatter reduces the image sharpness by creating a veil on the detector. It weakens contrast.

Scatter comes from the scattering of radiation by sidewalls and by the part itself. In the first case, it can be reduced with the help of accessories such as a diaphragm, a mask, or a screen. In the second case, it can also be reduced by anti-scatter grids, an air space between the part and the detector, or filters placed just behind the part.

For scatter coming from the part, appropriate settings can be made for radiation. To hinder the Compton effect, the primary beam can be hardened, which decreases the probability of the interactions responsible for this kind of scatter. But it is also possible to change domain, for example, by reducing the incident energy in order in order to remain in the field of the photoelectric domain.

6.2.4   Image quality indicators (IQI)

Image quality refers to the ability to make defect detection possible. At the least it depends on the parameters previously studied. To quantify it, a device composed of a series of progressive dimension elements is often used. For example, an *IQI system* (Fig. 6.18a) is based on a series of 19 wires of different diameters, divided into four groups (with an overlap) of seven wires. The material used is (light alloy) steel, copper, aluminum, and titanium.

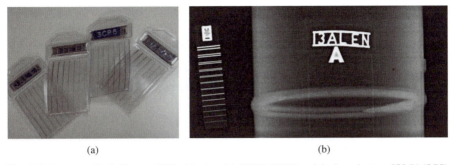

(a)   (b)

Fig. 6.18 Image quality indicators (IQI) with wires (a) (SOFRANEL) and duplex wire type IQI (b) (SGS)

The suitable IQI (with a material and diameter in accordance with the maximal dimension to be passed through) is put on top of the part, at its maximal thickness, and radiographed with the part. The EN ISO 19232 standard defines the wire number that has to be visible so that the image quality is correct and the radiography is accepted. The quality can be higher (+1, +2 etc.) according to the diameter of the wire that can be seen, and obviously it can be lower, too (−1, etc.).

Step and hole IQI work in the same way but are less frequently used, because they hide the part a little more than IQI with wires does.

The ISO 19232 standard defines a method for determining the total blur index of the image generated by the radiograms and radioscopic systems, with the help of a duplex wire type IQI (Fig. 6.18b). The pair of wires, whose image is the result of the fusion of the images of the two distinct wires, corresponds to the threshold of perceptibility.

The basic spatial resolution of the image is then deduced. These duplex wire type IQI are predominantly used in radioscopy, where the calculation of geometric blurring is more complex than in radiography.

6.2.5   Reference parts

The E446 (or E2422 for the digital) standard of the ASTM (American Society for Testing and Materials) provides a collection of radiographs presenting the different types of defects found in cast steel parts, with a thickness lower than 50 mm. These radiographs have been captured either with X-rays (250 kV), or with X- or γ-rays (1 MeV or Ir192, and 2 MeV or Co60). There are several types of defects with different levels of seriousness.

Importantly, the part will be declared "radiographically" acceptable if the size and number of defects are, at the most, equal to those defined in the standard (to the equivalent part surface).

Note that wire or duplex wire type IQI can in a certain way be considered as reference parts.

6.3   METHODS OF TESTING

6.3.1   Inspection protocols, time of exposure calculation

To control the process of image formation, it is necessary to know, in practice, the exposure rate of the X or γ beam that has passed through the thickness of a material.

   To do so, we have to know:

- The *exposure rate* at 1 m from the source, in the air. This depends on the source. Adjustment in kV and mA for X photons are refined, and the residual activity for the γ rays, according to the nature of the element (its period), and the initial activity is calculated.
- The source-detector distance. This depends on the environment, the size of the source, and the desired geometric blurring, in other words, on the required image quality.
- The thickness of the part.
- The attenuation law, which depends on the nature of the material and the defect being sought.

   For example, on a radiographic film, standard volume defects will give more or less dark indications, related to their attenuation coefficient and their dimension (Fig. 6.19). Porosity will be quite dark, an inclusion less dense than the material will be darker, and conversely. For example, an inclusion of tungsten in a TIG weld appears in white while the inclusion of slag appears dark.

   Non-volume defects (cracks) will give dark linear information on a film, if they are well orientated (parallel) to the radiation.

   Knowing the transmitted rate makes it possible to evaluate the safety conditions and determine integration time (digital) or exposure time (film), taking into account the detector's needs (its dynamic, for example), itself chosen with regard to the desired image quality.

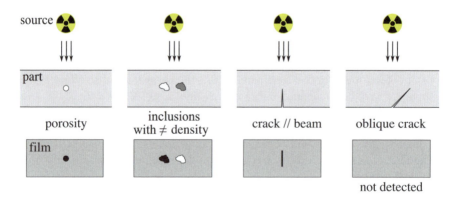

Fig. 6.19 Representation of volume or non-volume defects on analog film

6.3.2   Radiographed maximal thickness

Radiographed maximal thickness depends on the source, material, or adjustment used. Thickness sizes (in mm of steel) are in the order of:

- 80 to 100 mm for Ir192 sources
- 160 mm for Co60
- 100 mm for X tubes of 400 kV
- 200 mm for 1 MeV accelerators
- 500 mm for 5 MeV accelerators

Note that there is a 16 MeV accelerator that is used for the inspection of solid propellant in Ariane rockets. This material is suitable for neutron radiography, which is not available on sites, especially in Kourou.

6.3.3   Particular methods

Double-film technique
By using two films, one behind the other, the ***double-film*** technique reduces the time for carrying out the radiography by half. The two films are then adjusted one on top of the other when the inspection is visualized in order to combine their individual optical densities.

This technique is also used to radiograph parts with different thicknesses. In this simple case, when one of the two thicknesses is double the other, with one shot of X-ray or γ ray, two films are obtained that are separately inspected for the single thickness and adjusted for the double thickness. The shot is then saved but adjusting films is always a delicate operation. Digital image adjustment can be simplified with the help of suitable software.

Stereo-radiography
Radiography cannot inherently locate the depth of a defect, but stereo-radiography makes this possible (Fig. 6.20); by measuring the distance between the sources and the shifts on the film it becomes possible to calculate the depth position of the defect.

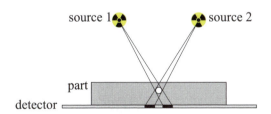

Fig. 6.20 Principle of stereo-radiography

6.3.4   1D, 2D, and 3D imaging modalities

The fluorescence of screens sensitive to X-rays creates a visible image of the object. In the past, this image was observed directly behind the screen, which is very dangerous for a non-protected observer. At the beginning of the 20th century, it caused numerous diseases among radiologists (for example, hand necrosis) and some of them even died.

Later, when this image was taken by a camera, it enabled observation at a safe distance from the dangerous radiation. A fluorescent screen portrayed the input side of a luminance amplifier, in which the electrons produced by X-rays at the surface of the screen were accelerated and turned into visible light, which provided a luminous image for the camera lens. This amplification strongly decreased the necessary radiation doses (this was the first procedure used for the inspection of luggage). The image on the screen could be recorded on magnetic tape, printed, or digitized.

Today, imaging modalities have changed depending on the purpose of the inspection. A linear detector array (single or multi-line, delayed or not, straight or bent) enables inspection of large dimensions (several meters) by associating numerous detection modules. This is called *radioscopy* (Fig. 6.22a) when the digital image is obtained by a detector and not via the digitization process of a radiographic film. Thus, the image is created by being associated with a particular movement. Note that appropriate image quality can be achieved in all cases, without necessarily being familiar with the laws of variation of parameters, which is essential in radiography. The different movements are the following:

- A translation perpendicular to the main direction of the linear detector array acquires the image column per column. That is the case for luggage checking systems in airports, which associates several linear detector arrays (often in a shape of a L so that the object can be seen under two incidences, which gives a better analysis).
- A rotation perpendicular to the main direction of the linear detector array enables the inspection of circular parts such as for the inspection of tires (Fig. 6.21).
- A rotation along the linear detector array gives an assembly of projections allowing the reconstruction of a section. As a matter of fact, for a non-homogeneous object, the local value of attenuation cannot be calculated directly from the attenuation obtained on a single photon path. A group of many (but limited) projections via an adapted reconstruction algorithm (initially via the Radon transform) can remove the indecision. This is known as *tomography* (or *scanner* in the medical field).

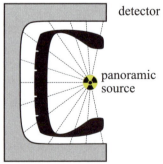

Fig. 6.21 A U-shaped linear detector for the inspection of tires (DETECTION TECHNOLOGY)

Tomography (Fig. 6.22c) is different from *tomosynthesis* (Fig. 6.22b) essentially through the fact that the image is rebuilt via a more limited number of views. In the medical field, this is of particular interest to the patient, because tomography is considered to leads to the absorption of a dose equivalent to 250 radiographs for an abdomen scan and to 1,000 radiographs for a heart scan! This is not an issue for industrial tomography: the choice between the two processes is made based on access to the part, as tomosynthesis experimentation is orthogonal to that of tomography.

The object can also be moved in a translation movement during the rotation in order to make a continuous acquisition in helicoidal geometry, which can be particularly useful for the inspection of tubes, bars, or process tomography.

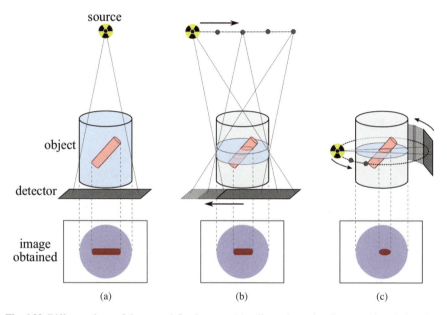

Fig. 6.22 Different views of the same defect between (a) radiography and radioscopy (the whole volume is projected), (b) tomosynthesis (reconstruction of a thick section), and (c) tomography (reconstruction of a thin section of a part).

Matrix detectors (2D) are often used. They differ according to the time rate, spatial resolution, detection technology, and the movement associated with them.
- There are permanent systems to perform simple 2D radiographies.
- Coupled to a rotation system, they enable tomographic reconstruction in 3D conical geometry (Fig. 6.23a). The rotation can continue after more than one rotation to rebuild successive temporal phases of a dynamic material (when moving or under stress): this is 4D imaging.
- Tomosynthesis systems (Fig. 6.22), or laminography, use complex movements to highlight a clear section of an object, while the rest remains blurred. Historically, laminography used to be carried out on film, with direct exploitation. Tomosynthesis (Fig. 6.23b) is carried out after, on digitized images.

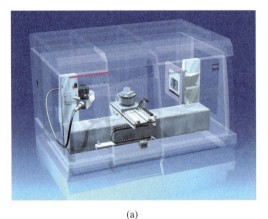 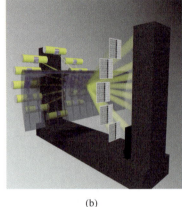

(a)   (b)

Fig. 6.23 Example of a 3D imaging cabin in conical geometry (a) (CARL ZEISSIM) and a laminography or tomosynthesis system (b) (YXLON)

6.4   EXAMPLES OF APPLICATIONS

The applications presented here show the richness of radiographic inspections: there are standard industrial cases and more complex expertise cases. Some situations are further explained in chapter 13 and 14 to demonstrate current NDT issues in the context of radiography and the solutions that have been reached or are under development (for example, the deterioration of information – signal and/or image – coming from the measurement environment, etc.).

6.4.1   Casting and forging

The identification of defects (Fig. 6.24) is simplified thanks to knowledge of the casting process (sand, metal, etc.). Radiographic testing in this case contributes more to the improvement of the manufacturing process by validating the arrangement of deadheads, vents, internal and external coolers, etc., than to welded joints.

6.4.2   Welding

Generally, the source-film relative positions (Fig. 6.25) have to respect simple rules: the film has to be positioned as closely as possible to the part (to reduce blurring), and the source has to generate a photon flux parallel to possible non-volume defects. In this way, they will be detected (Fig. 6.26).

In some cases, the arrangement of the inspection is more complex. The EN ISO 17636-1 standard shows a few examples of welded assemblies (Figs 6.27–6.28).

(a)   (b)

Fig. 6.24 Radiographs of shrinkage cavities on aluminum (a) (NTB) and of shrinkage cavities and blow-holes on cast iron (b) (LCND)

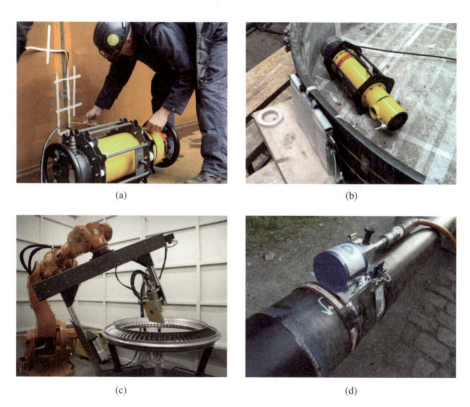

(a)   (b)

(c)   (d)

Fig. 6.25 Different assemblies of X radiography (a, b, c) (SOFRANEL) and gamma radiography (d) (GROUPE INSTITUT DE SOUDURE Gamma Prox patent)

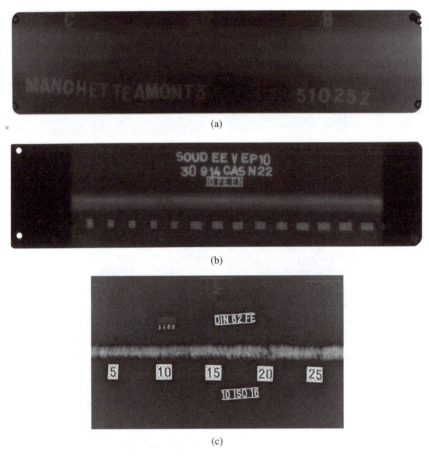

Fig. 6.26 Radiographs of welding with transverse cracks (a) (LCND), porosities (b) (LCND), blowholes, inclusions, and IQI (c) (GROUPE INSTITUT DE SOUDURE)

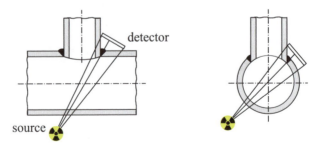

Fig. 6.27 Inspection arrangement for double wall, single image exposure of curved objects for an interpretation of the wall on the side of the film

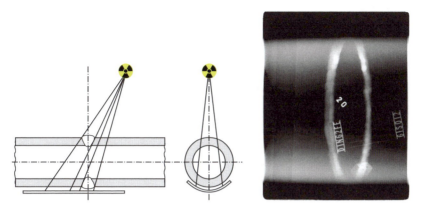

Fig. 6.28 Oblique inspection arrangement for ellipsoidal projection of the welding and the corresponding radiograph (SOFRANEL)

6.4.3   Metrology

Tangential radiography enables measurement of gaps in mechanical assemblies that are difficult to evaluate with other non-destructive techniques. Figure 6.29a shows a projection of the mechanical gap between two tubes and its measurement from the "line" profile of the obtained radioscopy.

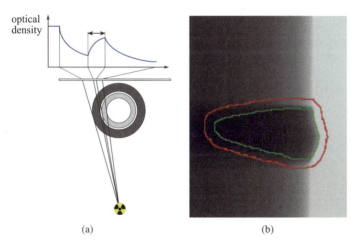

(a)   (b)

Fig. 6.29 Measurement of the mechanical gap between two tubes (a),automatic measurement of welding penetration (b) (LCND)

Figure 6.29b is the tangential radioscopy of a circular weld. Its contour (in green) is automatically found by image processing, by a B-Spline mathematical curve which, from an estimated initial contour (in red), finds the gradients of appropriate levels of gray.

Note that metrology dimensioning and inspection are increasingly carried out in 3D. The adjustment between the CAD model and the reconstructed part after a tomography highlights, for example, variations in dimension.

6.4.4 Tomography on composites

Tomography is now used industrially for the inspection of different materials and structures, including composites. Delaminations (Fig. 6.30) and several other defects, such as ply disorientation, fiber waviness, etc. can be identified. The interpretation of tomograms can be very complex and needs sophisticated image processing.

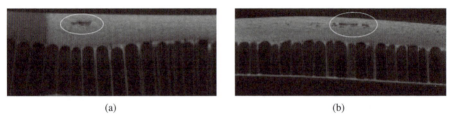

(a)   (b)

Fig. 6.30 Tomograms of delamination on composite material, top view (a), view from the left (b) (SAFRAN SNECMA)

6.4.5 Airport and safety

Monitoring systems in airports range from a simple cabin for hand luggage (Fig. 6.31) to security gates with linear accelerators for containers. Border control (luggage, cargo holds, containers, vehicles, etc.) requires heavy equipment to ensure the absence of hazardous (radioactive or explosive) substances, stowaways, etc.

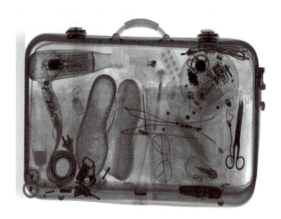

Fig. 6.31 Example of a luggage check (NTB)

6.4.6 Museums

National museums use X-ray imaging to inspect the inside of objects while protecting their integrity (sculptures, mummies, paleontology, mechanisms) or to reveal touch-ups in paintings (Fig. 6.32).

Fig. 6.32 Photograph and radiograph of the same painting (SOFRANEL)

6.4.7 Food-processing industry

In the food-processing industry, the aim is to detect foreign objects in the products: bone fragments in poultry fillets, glass shards in yoghurts or compotes, aluminum or Teflon fragments in cookies, etc. Radiography can even be used to sort types of fish according to the number of bones (Fig. 6.33a).

The agricultural industry also characterizes seeds (Figs 6.33b and c), sometimes for a strictly mechanical purpose, for example, in order to improve settings on sorting machines (blowers, vibrating conveyors, etc.).

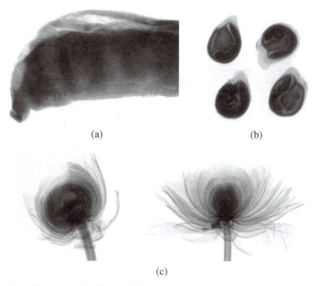

(a)   (b)

(c)

Fig. 6.33 Detection of bones in a fish fillet (a) (NTB), classification of tomato seeds (b) and flowers (c) (NTB)

As explained in section 2.6.4, the non-destructive evaluation in respect of the standard defining the proportion of foie gras bits compared to emulsion in a block of 1 kg pasteurized in a sealed metallic can is a complex issue. An X-ray tomography was suggested (Fig. 6.34), which clearly revealed the inside of the box and the differences in density between the two constituents.

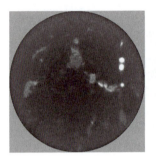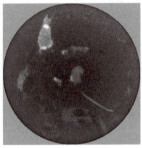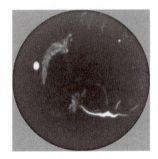

Fig. 6.34 Tomographic sections obtained at different heights or a can of foie gras showing whole foie gras in white (CTCPA, INSA LYON, LCND study)

6.4.8 Pharmacy

X-ray imaging techniques in the pharmaceutical industry are used for medicine production inspection such as the centering of active products in the coating of pills, the filling of each cell in the blisters, the correct centering of needles in syringes, the presence of foreign objects in opaque products (powders or liquids), the gear system of powder distribution boxes for inhalers, and so on. (Fig. 6.35).

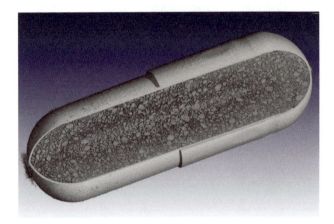

Fig. 6.35 3D tomographic reconstruction of a pill to inspect the density distribution of the active product (CT SCANNER)

6.4.9 Biology, medical applications, etc.

This book essentially introduces industrial NDT methods, but applications for the "living" world (biology, medical applications) are obviously numerous, whether in radiography (Fig. 6.36), radioscopy, tomography (for ionizing rays), or in echography (ultrasound), including special MRI techniques. These, however, will not be explained in this book.

 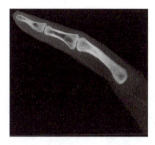 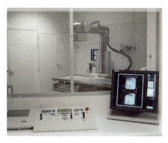

Fig. 6.36 Radiography of a mouse and a finger (NTB), medical installations

6.5   CURRENT AND FUTURE POTENTIAL

Radiography has strong positive factors for non-destructive inspection: interpretation is easy and contactless, there is a potential for automation without any major difficulty for production inspection (in radioscopy), and it is adaptable to 1D to 4D imaging protocols on a very wide range of spatial and temporal resolutions. Digital technologies for X-ray detection are well advanced.

Weaknesses in radiography are easy to identify:

- There are dangers related to ionization caused by photon-matter interactions, which need controlled radioprotection of facilities and operators.
- The testable material thickness is quickly limited because of physical loss. Much more powerful sources would have to be used, which are much too dangerous. For example, for photons at 100 keV, less than 1% of the radiation is transmitted through a 16 mm thick steel plate, and 10 cm for photons at 1 MeV.
- The detection of plane defects with a small opening is very delicate and depends on the orientation of the beam: only cracks parallel to the photon beam are detected and closed cracks are almost undetectable.
- Radiography is quite an expensive NDT method because on the one hand, two operators are required for safety, and on the other hand, no one is allowed to work in direct proximity, which is also a cost that has to be taken into account (zoning principles, § 6.7.3), and hence leading to the necessity for night shifts.

In industrial inspection, this is a technique that is chosen as "complementary" to ultrasonic inspection: manufacturing plane defects parallel to the surface are invisible to radiography, but accurately detected by ultrasound. In-service defects (fatigue), however, tend to be perpendicular to the surface, and in this case, radiography is favored.

Radiography (as a general inspection) is sometimes preferred to ultrasonic inspection (local) because inspection rapidity is an essential factor. This is the case when working under radiation, when operators have to carry out inspections as quickly as possible in order to limit the doses they are exposed to. But background interference caused by surrounding activity has to be tolerable.

The digitization of radiographic films improves the archiving process by reducing the volume of data. Nonetheless, it has to meet the standards defined in EN 14096-1. A normalized reference film composed of a series of targets is then used to monitor the performance of the digitization process, which can reach a spatial resolution of 25 μm or a 0.02 sensitivity to density contrast. A digitized film can also be the subject of numerous image processes (Fig. 6.37):

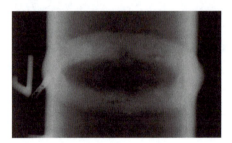 

Fig. 6.37 Digitized radiograph and application of 3D filter (ANAïS)

Improvements in radiography continue to advance, as outlined in the following sections.

6.5.1   Electronic tomography

This kind of imaging is based on the use of an electron beam, coming from an electronic microscope, and going through the sample. Typical resolutions of this kind of system are very low (a dozen nanometers approx.).

6.5.2   Quantitative imaging and reduction of artifacts

If detection and sizing of defects are simple undertakings in radiology, density measurement and material characterization are strongly disturbed by problems inherently related to the acquisition process: scattered radiation, artifacts, detector response, etc. 2D or 3D quantitative imaging is still challenging, especially when the measurement has to be accurate.

For more general applications, quantitative imaging is often used. It is sometimes called *radiometry*, for example, for material density analysis or pipe filling evaluation, etc.

Gamma absorption, gamma-densitometry, or *densitometry*, are also used in various applications, such as the characterization of wood density, which can be used to

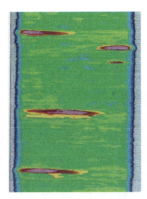

Fig. 6.38 Gamma-densitometry showing the density of a timber plank with knots (CEA DAMRI)

precisely date trees by detecting growth rings or locate knot areas, but also allows industrial cutting of quality timber planks while reducing scrap (Fig. 6.38).

6.5.3 Neutron radiography

Neutron radiography is an imaging technique that is very similar to X-ray radiography (attenuation, blurring, image quality, etc.) but the use of a *gadolinium* converter is needed because neutrons do not directly ionize the film.

The used sources are nuclear reactors, particles accelerators, and a few isotopic sources (like californium), but the fundamental difference with radiography is based on the fact that the incident particles are neutrons and that they interact with matter in a different way, that is, they do not follow a logical path. Thus, hydrogenated products attenuate significantly as opposed to metallic materials (Fig. 6.39).

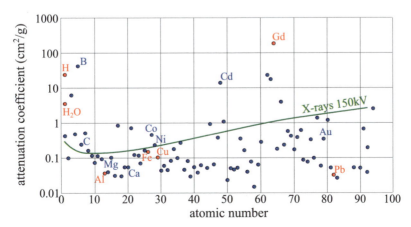

Fig. 6.39 Attenuation coefficients for X-rays (150 kV, in green) and thermal neutrons (blue points, some materials are highlighted in red)

This technique has become a complementary form of imaging to X-radiography (Fig. 6.40), notably for the detection of hydrogenated products (very high attenuation) in any materials.

Fig. 6.40 Complementary methods: radiography, photography, and neutron radiography of a calculator (CEA)

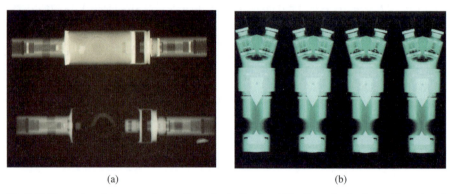

(a)   (b)

Fig. 6.41 Comparison between neutron radiography (bottom) and X radiography (top) of a delay relay of an Ariane rocket (a) (CEA) and neutron radiography of pyrotechnics valves (b) (CEA)

The use of neutron radiography is of particular interest for pyrotechnics inspections (Fig. 6.41). Container filling, explosive powders, and detonating cord compactness have to be perfectly synchronized in order for ejection seats and jettison canopies to function correctly or rockets and missile modules to separate.

Neutron radiography has also significant potential when it comes to the inspection of sealing joints, greasing, bonding, etc.

The matter activation (when it becomes radioactive) after neutron inspection prevents it from being more widely applied: depending on the inspected material, a simple quarantine lasting a few hours or days is enough but for others, the part becomes useless and is wasted.

6.5.4   Synchrotron

The use of radiation sources with very high luminance enables near monochromatic sources (quantitative imaging, photon phase imaging, etc.) to be obtained while still maintaining an considerable flow of particles. It also allows the observation of dynamic phenomena.

6.5.5   Crystallography

The elastic scattering of monochromatic X-rays (in small angles "SAXS" or wide angles "WAXS") is reflected by diffraction figures that are typical of the form and size of the molecules and of the inter-atomic distances in crystal structures.

6.5.6   Flash radiography

A visualization of ultra-fast phenomena (explosions) is possible thanks to synchronized systems based on the emission of X-rays by impulsions of a few nanoseconds that are very intense.

6.5.7   Secondary radiation imaging

Fluorescent and/or scatter radiation gives further information, complementary to X radiography, about chemical composition and/or density. Specific collimation processes are necessary for the creation of the image.

For example, some full-body scanners used in airports for at-risk flights are based on the observation of low-energy Compton backscatter radiation, which allows an inspection without radiation filtering through the body.

6.6   RADIOPROTECTION

6.6.1   Charter of good practice in industrial radiography

In 2010, the French Nuclear Safety Authority (ASN Autorité de Sûreté Nucléaire) with the following French organizations: Directions régionales des entreprises, de la concurrence, de la consommation, du travail et de l'emploi (DIRECCTE), the Caisses régionales de l'assurance maladie (CARSAT formerly CRAM) of the Auvergne Rhône-Alpes region, and professionals within this sector have contributed to the creation of a charter of good practice in industrial radiography.

Because of the high activity of the implemented radioactive sources, industrial radiography represents a risk for workers, and it is necessary to limit and prevent this risk. The purpose of this procedure is to make the application of rules easier, to change practices and habits in order to improve prevention as well as the intervention conditions of workers, and to reduce the doses recorded by technicians in industrial radiography.

The charter clearly introduces good practice in industrial radiography in order to help professionals better understand and integrate the general principles of prevention at each step of a project. The charter is based on the following points: the reduction of worker exposure regarding the principle of justification (i.e., exposure must be justified by its advantages as opposed to the risks inherent to it), the preferred use of NDT technologies presenting less risks for the health of workers, the principle of optimization (the exposure has to be reduced to the lowest level reasonably possible), and the principle of limitation (the exposure has to be under the regulation threshold).

6.6.2 Principle of justification in radiography

The major disadvantage in industrial radiography (γ and X radiation) is that operators (even the public) are subjected to a major risk of external exposure. Every exposure must be justified by the advantages related to the risks; the use of technologies presenting less risk to the health of workers has to be favored.

This is the case for ultrasound, in TOFD version (§ 7.3.5), and other NDT techniques that have a lower risk for workers' health and can replace, in respect to the conditions, industrial radiography. But if radiography is chosen, the use of X radiography (X-ray generator) will then be favored as opposed to γ radiography (gammagraphy).

6.6.3 Principle of optimization and limitation: zoning, dosimetry

The principle of optimization drives a dosimetry follow up in real time. In order to choose the right dosimeter, other parameters have to be taken into account, such as the dose rate. If it is low and not subjected to variations, a passive dosimetry is acceptable (low investment, the use of a film is sufficient). But if there is risk of large variations and a high dose rate, the use of an active dosimeter with continuous display of the dose and/or dose rate with visual and sound alarms is recommended. The idea is that the worker will understand and gauge the level of risk and so limit his or her exposure.

According to the same principle, in order to evaluate the levels of exposure, we have to determine the nature and scale of the risk due to ionizing radiation, according to the characteristics of the sources (sealed, unsealed, type and energy of the emitted radiation, dose rate, emission duration), installations (collective protections or not), etc. All of this has to be considered for the strictest normal conditions of use. In France zoning is formalized in accordance with the erections of panels that comply with Appendix I from the decree of May 15, 2006, in respect of the colors mentioned in Figure 6.42.

For example, on the borderlines of monitored or controlled zones, the effective dose likely to be absorbed within a month by any worker is lower than 0.080 mSv, which is the established value on a yearly basis of 1 mSv for a non-exposed worker and converted to a month.

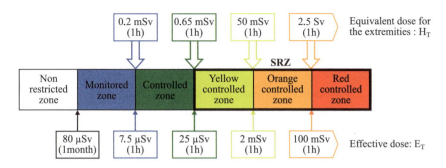

Fig. 6.42 Definition of the restricted zones (RZ, blue, green) and special restricted zones (SRZ, yellow, orange, red). Fixed Installations. Dose values H_T and E_T are integrated over the considered period.

In France, authorized workers have to be in the A category (Radiation workers directly affected by radiation work – in French: DATR *Directement Affectés aux Travaux sous Rayonnements*), with a maximal effective dose of 20 mSv/year in controlled zones. And in the B category (Non-Directly Affected by Radiation Work – in French: NDATR *Non Directement Affectés aux Travaux sous Rayonnements*) with a maximal effective dose of 6 mSv/year in the monitored zones. Beyond these limitations, the public is concerned, and is not specifically followed up (no dosimetry, etc.), and the maximal effective dose is limited to 1 mSv/year.

6.7   REFERENCES

ASN, CRAM, DIRECTION DU TRAVAIL, *Charte de bonnes pratiques en radiographie industrielle*, 2010.

ASTM, *Standard Reference Radiographs for Steel Castings Up to 2 in. (51 mm) in Thickness*, ASTM E446-98, 1998.

BUSHBERG J.T., *The Essential Physics of Medical Imaging*, WOLTERS KLUWER, 2011.

COFREND, *Démarche de justification de la radiographie gamma*, Les cahiers techniques de la Cofrend, Editions Lexitis, 2012.

DANCE D.R., CHRISTOFIDES S., MAIDMENTA. D.A., McLEAN I.D., NG K.H., *Diagnostic Radiology Physics*, Handbook of IAEA, 2014.

DILLENSEGER J.P., MOERSCHEL E., *Guide des technologies de l'imagerie médicale et de la radio-thérapie: Quand la théorie éclaire la pratique*, Editions Masson, 2009.

FANET H., *Imagerie médicale à base de photons: Radiologie, tomographie X, tomographie gamma et positons, imagerie optique*, Hermes Science Publications, 2010.

GRANGEAT P., *La tomographie: fondements mathématiques, imagerie microscopique et imagerie industrielle*, Hermes Science Publications, 2002.

GUEUDRE C., MOYSAN J., CORNELOUP G., Geometric Characterization of a Circumferential Seam by Automatic Segmentation of Digitized Radioscopic Images, *NDT&E International*, Vol. 30, no. 5, 1997: 279–285.

GUEUDRE C., MOYSAN J., CORNELOUP G., Radioscopic Images Segmentation by "Edge and Area" Combined Approach for Weld Geometric Characterization, *Research in Nondestructive Evaluation*, 12, 2000: 179–189.

HUSSEIN E.M., *Handbook on Radiation Probing, Gauging, Imaging and Analysis: Volume I: Basics and Techniques (Non-Destructive Evaluation)*, Kluwer Academic Publishers, 2003.

INSAVALOR, *Contrôle Non Destructif par Radiographie niveaux 2 et 3*, Lecture documents, 2014.

KNOLL G.F., *Radiation Detection and Measurement*, 4th Edition Wiley, 2010.

LAVASTRE D., BOUVET P., Guide pour la radiographie numérique des pièces moulées, COFREND Congress, 2011.

MOYSAN J., GUEUDRE C., MOULINEC H., Reconstruction tomographique 3D d'une soudure circulaire à partir de projections partielles issues de radioscopies tangentielles, *Revue Traitement du Signal*, Vol. 17, no. 2, 2000: 113–123.

NF EN 12681, *Fonderie Contrôle par radiographie*, 2003.

NF EN ISO 11699-1-2, *Essais non destructifs, Films utilisés en radiographie industrielle*, 2012.

NF EN ISO 19232-1-2-3-4-5, *Essais non destructifs, Qualité d'image des radiogrammes*, 2013.

NF EN ISO 17636-1-2, *Contrôle non destructif des assemblages soudés, Contrôle par radiographie*, 2013.

PETERZOL A., LÉTANG J. M., BABOT D., A Beam Stop Based Correction Procedure for High Spatial Frequency Scatter in Industrial Cone-Beam X-Ray CT, *Nuclear Instruments and Methods in Physics Research Section B*: Beam Interactions with Materials and Atoms, Vol. 266, 2008: 4042–4054.

RUAULT P.A., *Radiologie Industrielle, tomes 1 et 2*, Institut de Soudure, Publications de la Soudure Autogène, 1991.

VUILLEZ J.P., *Radioprotection*, UE3-1 Biophysics course, Université Joseph Fourier, 2011.

WILS P., LÉTANG J. M., BRUANDET J. P., Secondary Radiations in Cone-Beam Computed Tomography: Simulation Study, *Journal of Electronic Imaging*, Vol. 21, no. 2, 2012.

CHAPTER 7

ULTRASONIC TESTING

Ultrasonic waves are mechanical waves that propagate through all materials (solid, liquid, and gaseous). The frequencies are higher than those of sounds and range from 1 to 10 MHz in industrial inspections. These waves are often generated by a piezoelectric transducer that turns an electric impulsion into vibrations. By liquid coupling the transducer to the part, in ideal conditions, ultrasonic waves propagate in a straight line at constant velocity (or celerity), until they meet an interface between two media. The energy is partially reflected and transmitted by this macroscopic interface (surface of the part, crack, inclusion, porosity, etc.) or microscopic interface (microstructure, grain boundary, etc.). The receiving transducer (which can be the emitter or another one), allows, via a device, echoes that are characteristic of ultrasonic propagation to be visualized. Measuring the reflected and transmitted amplitudes, according to the reflecting surface of the reflector (therefore its size) and the time of flight make it possible to determine the distance from the reflector.

Ultrasonic testing has major benefits such as ease of implementation, non-necessary accessibility to the two sides of the part, good adaptation to the natural orientations of most defects, the possibility of passing through great thicknesses, a direct link to the mechanical characteristics of the material, and ease of information digitization.

Disadvantages essentially come from the difficulty in transmitting the wave from the transducer to the part, which imposes the use of coupling (usually liquid), electro-magnetic transducers, or lasers. The very high sensitivity of propagation to the rates

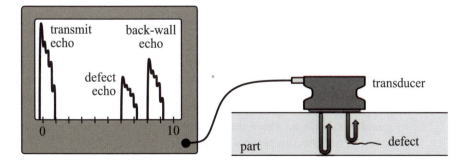

Fig. 7.1 Principle of ultrasound testing and representation of echoes

of heterogeneity or anisotropy of the material or to the environment (temperature, state of stress, etc.), is also an inconvenience in industrial inspections. But it is also thanks to these factors that the measurement of one parameter is possible; this proves that ultrasonic testing is, without a doubt, the most efficient method of non-destructive characterization of materials and structures.

7.1   PHYSICAL PRINCIPLES

7.1.1   Ultrasound

The frequency range of sound waves is from hertz to gigahertz. (Audible) *sounds* go from 15 Hz to 15–20 kHz, and *ultrasound*, inaudible for humans, covers the field from 20 kHz. In industrial inspection, frequencies from 1 to 10 MHz are usually used. For frequencies higher than 50 MHz, we talk about acoustic microscopy.

An *ultrasonic wave* refers to the vibrating displacement of a medium that leads to local pressure changes in the case of fluids or stresses in the case of solids. The measurement or observation of the propagation of these variations allows a better understanding of the medium, and it is on this principle that ultrasonic testing is based. Testing is non-destructive because the vibration amplitude is small.

7.1.2   The propagation of ultrasonic waves in perfect (ELHI) media

Propagation models used generally assume that waves are plane (the wavefronts of a *plane wave* are theoretically infinite planes) and *monochromatic* (one single frequency). The first hypothesis is accessible, but the *impulse mode* (a wide range of frequencies) is often favored in NDT: the associated temporal signal is short and gives a temporal resolution (which will in practical terms better locate a defect in depth). So, even if the theory is not respected, its study helps us understand the main tendencies of ultrasound behavior.

The displacements caused by wave propagation in matter change the state of the interatomic bonds at a specific moment. If, by hypothesis, the nature of these bonds was perfectly elastic, the wave would propagate in the medium without using energy, deforming or deviating, and with a celerity value independent from frequency but only depending on the material and the type of wave.

If we consider a homogeneous medium subjected to an ultrasonic perturbation, the hypothesis of small perturbations of the theory of elasticity can be applied. Stresses are then linearly linked to deformations according to *Hooke's law*, and the *equation of propagation* of an elastic wave can be written.

The solution to this equation of propagation shows that, in general cases, three plane waves can propagate with different speeds and orthogonal polarizations:
  • one *quasi-longitudinal or pressure* wave (QP),
  • two *quasi-transverse or shear* waves: one with a *horizontal polarization* (QSH) and the other with a *vertical polarization* (QSV).

Note that only longitudinal polarization is possible in fluids.

In the particular case of a "perfect" solid medium of density ρ, called **ELHI** (elastic, linear, homogeneous, and isotropic), the two transverse waves merge. The celerities of the two remaining types of waves (Fig. 7.2), *pure longitudinal* (the polarization is parallel to the displacement: this is called a *pressure or compressional wave*) and *pure transverse* (the polarization is perpendicular to the displacement: this is called a *shear wave*), are linked to the elastic mechanical parameters of the solid (Young's modulus E and Poisson's ratio v) by

$$C_P = \sqrt{\frac{E(1-v)}{\rho(1+v)(1+2v)}} \quad \text{and} \quad C_S = \sqrt{\frac{E}{2\rho(1+v)}}.$$

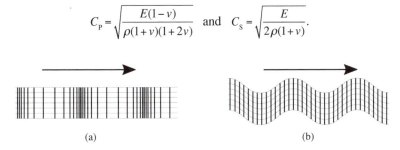

(a)   (b)

Fig. 7.2 The two pure modes: pressure and shear waves

In fluids, where the longitudinal wave is the only one that exists, celerity is linked to the module of compression K for the fluid of density ρ, with the relation

$$C_P = \sqrt{\frac{K}{\rho}}.$$

Thus, for a material considered "perfect" and of infinite dimensions, it is possible to calculate the celerity of the waves called *bulk waves* (Table 7.3). Note that the celerity of transverse waves is always lower than that of longitudinal waves.

Table 7.3 Ultrasonic celerity (considered to be perfect) in different industrial media

	density $(kg \cdot m^{-3})$	P wave celerity $(m \cdot s^{-1})$	S wave celerity $(m \cdot s^{-1})$
Metallic Construction Steel	7800	5920	3250
Industrial Aluminum	2700	6300	3080
Industrial Copper	8930	4600	2260
Water	1000	1480	–
Air	1.2	330	–
Glycerin	1260	1920	–
Plexiglas	1200	2730	1430
Lead	11370	2400	700
Titanium	4500	5990	3120

7.1.3   Ultrasonic wave propagation in real media

The movement of a monochromatic plane wave in an ordinary environment (not considered "ELHI" anymore), which can be inelastic, non-linear elastic, heterogeneous, and/or anisotropic, will be influenced by these parameters. This shows that it is sometimes an advantage (possible access to parameter measurements), but its application in NDT is often limited: ultrasound can be attenuated, slowed down or deflected, which will make the detection, location, characterization, and/or the sizing of the defects difficult.

Inelastic medium

A perfectly elastic medium subjected to a stress deforms without internal friction and goes back to its initial form when the stress stops, without using energy. If the medium is *inelastic*, or *viscoelastic* (plastic materials, resin, etc.), part of the energy of the ultrasonic wave is turned into heat (see the thermographic image in Figure 7.4 obtained after the passage of an ultrasonic wave). This is *attenuation by absorption*, which is relatively low in non-destructive testing of metallic materials.

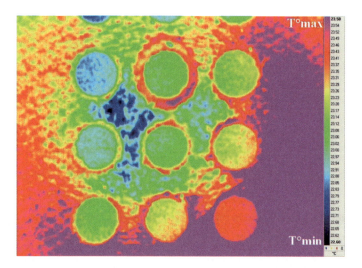

Fig. 7.4 Sonic thermogram of a drilled carbon composite part, with the apparition of delamination caused by drilling (MINES DOUAI)

Non-linear elastic medium

The elastic behavior of a medium is considered linear if each displaced atom, according to its position of equilibrium, complies with Hooke's law. In reality, because of the non-linearity of the interatomic force-distance relation, elasticity constants of a higher level have to be taken into account. Ultrasonic celerity (P and S) depends on the state of stress σ of the material:

$$C_P^\sigma = C_P^0(1 + A_P\sigma) \quad \text{and} \quad C_S^\sigma = C_S^0(1 + A_S\sigma).$$

A_P and A_S refer to the constants of acoustic-elasticity linked to the elasticity constants of the material.

A solid, initially isotropic, becomes anisotropic in a stressed state, and the characteristics of this anisotropy depend on the applied stress. The elastic non-linearity of an environment is often disregarded during ultrasonic testing for defects. But it can be used to estimate the applied or residual stresses in a material, for example, to evaluate the stress caused by the tightening of steel bolts (§ 7.4.8).

Non-linear acoustics (NLA) has only been explored in recent years. Paragraph 7.5.1 introduces, in perspective, the current work that has been done, showing the great potential of this new direction for research.

Anisotropic elastic medium
The behavior of an anisotropic elastic solid changes according to the orientation of the loads in relation to the material. This elastic anisotropy is found in single crystals and in polycrystalline materials textured by a particular manufacturing process (*texture* refers to the orientation of the grains in a preferential direction bringing out their anisotropic characteristics), such as rolling, forging, or some forms of welding, or in most composite materials.

Anisotropy causes, for each propagation direction *n* (normal to the front wave), the propagation of three non-pure plane waves QP, QSV, and QSH (§ 7.1.2), whose polarizations are no longer strictly parallel and perpendicular to the displacement. Each wave is defined by a *phase velocity* V_p (parallel to *n*) and an *energy velocity* V_e (parallel to the energy flux), which can be considered as the "real" direction of propagation (Fig. 7.5) but is no longer parallel to *n*.

So, waves do not propagate according to *n*, as in isotropic media. An expected consequence concerns the difficulty in properly detecting and locating a defect. The reflector will logically be normal to the wave front (isotropic material hypothesis), while it occurs in the energy direction, which is unknown when the anisotropy is unknown.

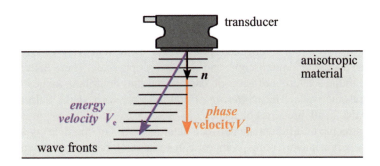

Fig. 7.5 Phase (V_p) and energy (V_e) velocities

Energy velocity is always perpendicular to *slowness surface*, which represents the opposite of velocity for all angles of propagation. Slowness surface for an isotropic environment is a circle. Figure 7.6 shows the case of two interfaces: isotropic-isotropic (Fig. 7.6a) and isotropic-anisotropic (Fig. 7.6b).

These curves graphically show the propagation direction of the transmitted waves. First, the vertical line given by the intersection of the reflected wave and the slowness curve of the incident medium is used. This vertical line crosses each slowness surface of the transmission medium at a point. The two local perpendiculars to the slowness surfaces help find the directions of propagation of the transmitted waves (notion of *acoustic ray* directly linked to the propagation of energy).

In the first case, for a pure P wave incident in a fluid, there are two transmitted pure P and S waves in an isotropic solid. In the second case, for a pure P wave incident in an isotropic steel (base metal, for example), two waves QP and QSV (n and V_e are not collinear) are transmitted into the weld.

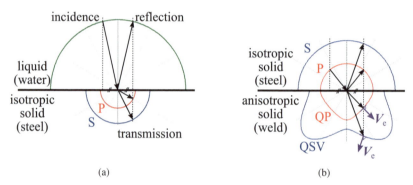

(a)   (b)

Fig. 7.6 Slow surfaces (a) between a fluid and an isotropic environment (b) between an austenitic steel and a weld

Acoustic radiation is in the plane of incidence only if this latter crosses the slowness surface according to one of its planes of symmetry. If this is not the case, a 3D representation of the slowness surfaces and polarizations will be necessary to understand the possibilities of propagation.

Heterogeneous elastic medium

The elastic behavior of a homogeneous medium is identical in every way. But heterogeneity can appear at a microscopic and/or macroscopic scale. These heterogeneities contribute to the scattering of acoustic energy, and thus contribute to the attenuation of the ultrasonic wave. The *scattering* of an ultrasonic wave is the reflection of an incident wave in all directions of space.

It is possible to differentiate between heterogeneities of "particle" type, such as second phase precipitates, pores, composite material reinforcement or concrete aggregates, and heterogeneities linked to the polycrystalline nature of the metallic materials (grains/aggregates).

Remember that a medium can always be "homogenized" through ultrasound, by choosing a work frequency lower than the value related to the size of heterogeneities. But this frequency is likely to be too low for an accurate inspection of the matter.

Attenuation: the case of polycrystalline materials

Generally, the theoretic displacement according to the x axis of a monochromatic plane wave in an infinite ordinary medium can be written as:

$$u(x, t) = U_0 \, \boldsymbol{P} \, e^{-\boldsymbol{a} \cdot \boldsymbol{x}} \, e^{(i(\boldsymbol{k} \cdot \boldsymbol{x} - \omega t))}$$

with: U_0 initial amplitude
      $\boldsymbol{k} = \omega / C \, \boldsymbol{n}$ wave vector
                 with ω angular frequency
                            C wave celerity
                            $\boldsymbol{n}$ normal to the wave front
      $\boldsymbol{P}$ polarization vector
      α global attenuation coefficient (absorption and scattering), which is a vector when it is linked to the direction.

In this theoretical case, absorption is negligible, and grain scattering is all that remains. Grain orientation is never random, but totally deterministic, with a law that is not often known. If the law is known, propagation can be predicted and modeled in a more realistic way. Chapter 14 describes a predictive model of the solidification of a weld (MINA) which makes the simulation of ultrasonic propagation possible.

When the distribution of the orientation of the grains is "regular," the structure is globally isotropic or macro-isotropic, even if each grain has the anisotropic properties of its crystalline structure. This causes attenuation by scattering, which takes into account the intrinsic anisotropy of each crystal.

In the particular case of a single phase structure with equiaxed grains, almost spherical and with appropriate anisotropy (not too high), some models have been suggested that take into account the importance of grain size according to the wavelength λ (with $\lambda = C/f$).

Three *scattering domains* are defined (note that some research is trying to develop "unified" theories), and the coefficient of ***attenuation by scattering*** α_{d} is given by:

- Rayleigh domain   for $\lambda \gg 2\pi\bar{d}$ $\alpha_{\mathrm{d}} = A_1 \bar{d}^3 f^4$
- Stochastic domain   for $\lambda \approx 2\pi\bar{d}$ $\alpha_{\mathrm{d}} = A_2 \bar{d} f^2$
- High frequency domain   for $\lambda \ll 2\pi\bar{d}$ $\alpha_{\mathrm{d}} = A_3/\bar{d}$

with $\bar{d}$ =   average diameter of the grains
        A_1, A_2, A_3 =   constants depending on the material
        f =   frequency.

When the grains are small compared to the wavelength, the wave only sees a homogeneous structure of grains. In the opposite case, ***structural noise*** is created due to reflection-transmission phenomena at the level of grain boundaries. Chapters 13 and 14 give further information about these complex phenomena involving scattering, attenuation, and structural noise.

7.1.4 Reflection-transmission to the interfaces (perfect media), acoustic impedance

The theoretical case of an infinite propagation medium does not exist in industrial inspection. The medium is limited by *interfaces* which separate two media with different *acoustic impedances Z*, that is, the ability to resist the displacement imposed by the wave:

$$Z_P = \rho C_P \qquad Z_S = \rho C_S.$$

Direction of propagation, oblique and normal incidence
Generally, the oblique incidence of a longitudinal plane wave to a plane interface between two isotropic and elastic solids creates a phenomenon of reflection and transmission of the wave with *conversion of mode*. So, four different waves are generated (Figs 7.7 and 7.8) which correspond to two possible modes in each solid.

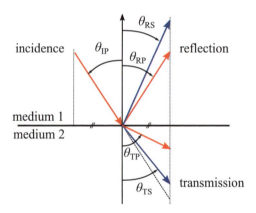

Fig. 7.7 Directions of propagation at an interface

The conditions of the displacement (or velocity) continuity of particles and stress continuity when a border separating two environments is crossed are indications giving the directions of propagation of each of these waves.

They are linked by the *Snell-Descartes law*:

$$\frac{sin(\theta_{IP})}{C_{P1}} = \frac{sin(\theta_{RP})}{C_{P1}} = \frac{sin(\theta_{RS})}{C_{S1}} = \frac{sin(\theta_{TP})}{C_{P2}} = \frac{sin(\theta_{TS})}{C_{S2}}.$$

We can easily show that: $\theta_{IP} = \theta_{RP}$ (we would have $\theta_{IS} = \theta_{RS}$ in the case of an incident S wave).

In the case of normal incidence of a plane longitudinal wave to an interface, the law of Snell-Descartes shows that the reflected L wave and the transmitted L wave are also normal to the interface.

Critical angles, Rayleigh waves

The first *critical angle* refers to the disappearance at the surface of the transmitted longitudinal wave (sin $\theta_{TP} = 1$), that is, an angle $\theta_{IP} = \arcsin (C_{P1}/C_{P2})$. The second critical angle refers to the disappearance at the surface of the transmitted transverse wave (*sin $\theta_{TS} = 1$*), that is, an angle $\theta_{IP} = \arcsin (C_{P1}/C_{S2})$. In this case, the theoretical reflection is total. Figure 7.8 represents the two cases:

A *surface wave* is created in medium 2, of $C_{SW2} \approx 0{,}9\ C_{S2}$ speed, slightly (a few degrees) past the second critical angle. These waves were discovered by Lord Rayleigh and named after him. When the second medium is not a vacuum, we talk about generalized *Rayleigh waves*.

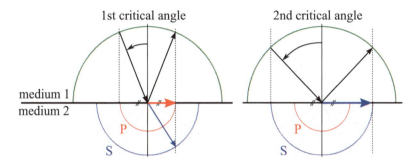

Fig. 7.8 Existence of 2 critical angles in acoustics

In a perfect medium, Rayleigh waves are non-dispersive; they do not depend on the frequency but only on the angle of incidence. These surface waves have energy to a depth that is in the order of that of the wavelength, and are essentially used to detect the presence of emerging defects.

Paragraphs 7.1.5 and 7.1.6 will deal with their modification in the case of propagation in a part that has a thickness in the order of that of the wavelength.

Reflection-transmission coefficients at an interface: normal incidence

The expression of the conditions of continuity is also an indication for calculating the amplitudes of the displacement at the passage of each reflected and transmitted wave. In the simple case of normal incidence, with A_{RP} and A_{TP} the reflection and transmission coefficients in amplitude of displacement and for an incident longitudinal wave P going from medium 1 to medium 2, the expression is

$$A_{RP} = \frac{\text{"1to2" reflected amplitude}}{\text{incident amplitude}} = \frac{Z_1 - Z_2}{Z_1 + Z_2}$$

$$A_{TP} = \frac{\text{"1to2" transmitted amplitude}}{\text{incident amplitude}} = \frac{2Z_1}{Z_1 + Z_2}$$

with $1 + A_{RP} = A_{TP}$ (incident + reflected amplitude = transmitted amplitude).

If we think in terms of energy, the energy coefficients E_{RP} and E_{TP} are

$$E_{RP} = \frac{\text{"1to2" reflected energy}}{\text{incident energy}} = \frac{(Z_1 - Z_2)^2}{(Z_1 + Z_2)^2}$$

$$E_{TP} = \frac{\text{"1to2" transmitted energy}}{\text{incident energy}} = \frac{4Z_1 Z_2}{(Z_1 + Z_2)^2}$$

with $E_{RP} + E_{TP} = 1$ (conservation of mechanical energy).

The higher the impedance of a medium is, the higher the energy necessary to move it. It is the contrast of impedance at the interface that characterizes the distribution of energy. Thus, the energy reflected at the material-air interface is maximal (bottom of a part, crack, etc.). Calculations show that in immersion testing (in water) of a steel part, only 12% of the energy is transmitted (88% of it is then reflected) at the first interface (at the input in the part) when performing the test.

Coefficients of amplitudes of displacement should be used, because they are directly proportional to the electric amplitudes (volt) recorded by the piezoelectric system (§ 7.2). Coefficients of energy are more suitable for a qualitative approach.

Reflection-transmission coefficients at an interface: oblique incidence

In the case of oblique incidence, the values of energy and amplitude coefficients for transmission and reflection are more complex. Figure 7.9 shows the energy coefficients of reflection and transmission E_{TP} and E_{TS} for a Perspex-aluminum interface.

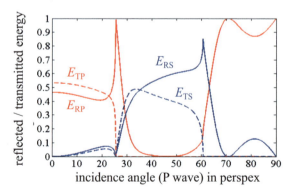

Fig. 7.9 Energy coefficients for reflection (continuous line) and transmission (dotted line) for a Perspex-aluminum interface (red line: P wave, blue: S wave)

We can see that the energy values calculated for the normal incidence ($E_{RP} = 0.46$ and $E_{TP} = 0.54$) are different from those calculated in amplitude ($A_{RP} = -0.68$ and $A_{TP} = 0.32$). The first critical angle (disappearance of the refracted pressure wave) appears at 26° and the second critical angle at 61°. We find that the highest amplitude, transmitted in aluminum, is the TS (transmitted shear) wave obtained with a 32° incidence.

Snell's law would show that we are close to the "S45 wave" (transmitted shear wave propagating at 45°), generally used in NDT, because it is dynamic and propagates alone in the material (there is no possible confusing diagnosis with the presence of TP in the case of a lower incidence).

7.1.5 The case of a plate with finite thickness: resonance

In this case, the results depend on the media, the part thickness, which is characterized by two interfaces, and the wave frequency. For a normal incidence (Fig. 7.10), when the thickness of a part is equal to the wavelength, *resonance* phenomenon occurs.

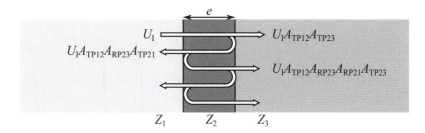

Fig. 7.10 Transmission through a plate

Thus, in the case of transmission through a plate with parallel faces, in normal incidence, the total transmission (i.e., the accumulation of outgoing waves) is calculated with the coefficient of energy:

$$E_{TP} = \frac{"1\,to\,3"\ transmitted\ E}{incident\ E} = \frac{4Z_1 Z_3}{(Z_1 + Z_3)^2 \cos^2 2\pi e / \lambda_2 + (Z_2 + Z_1 Z_3 / Z_2)^2 \sin^2 2\pi e / \lambda_2}$$

which points out 2 particular cases:

$$E_{TP} = \frac{4Z_1 Z_2}{(Z_1 + Z_2)^2} \quad \text{for } e = n\, \lambda_2/2 \text{ (maximal transmission, the plate is "non-existent")}$$

$$E_{TP} = 1 \text{ for } e = (2n + 1)\, \lambda_2/4 \quad \text{if } Z_2 = (Z_1 Z_3)^{1/2} \text{ (total transmission)}.$$

This latter case is used by transducer manufacturers to calculate the thickness of the protective workface on the front side (attached to the sensor), thus maximizing the transmission of ultrasound according to the inspected material (*impedance match*).

For oblique incidence, the results will depend on the calculation of the outgoing elements from the plates:

- Either we consider the first direct paths (first rays), because it is possible to differentiate between them, as in pulse testing. This requires the application of Snell's law, but the thickness of the part is not involved.
- Or all the waves coming out of the part are gathered, as in wave train (or packet) testing. In this case, the coefficient of global transmission is used, by integrating possible phasing according to the thickness of the plate and the wavelength.

For example, the coefficients of transmission for a stainless steel plate immersed in water are shown in Figure 7.11. When the amplitudes are cumulated, the thickness of the plate (2 mm here) and the frequency (1 MHz here) must be known.

On the left, we can say that the highest amplitude transmitted in steel is the S wave obtained with an incidence of 16°. Snell's law would show that we are close to the S45. On the right, the maxima are theoretical cases of total transmission of the amplitude. The peaks refer to incidence angles in water (9.2°, 17.8°, and 33.6°) for which Lamb waves are generated in the plate (cf. § 7.1.6).

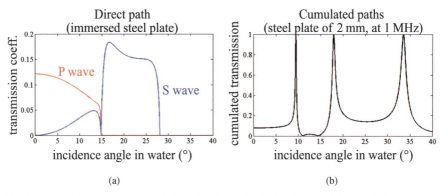

Fig. 7.11 Coefficients of direct transmission (a) and cumulated (b) for 2 mm stainless steel plate immersed in water at a frequency of 1 MHz

7.1.6   Guided waves and Lamb waves

The presence of an inclined plane wave in a medium limited by two parallel planes causes the *guided waves* phenomenon, resulting from multiple reflections. Under some geometrical conditions, it is possible to obtain "in phase" waves reflecting on the interfaces, creating guided waves, also known as *Lamb waves* (named after their discoverer), in the case of plates surrounded by vacuum.

By bringing parallel plane interfaces into the equations of propagation of plane waves in an ELHI infinite perfect medium, we obtain a *dispersion equation* of Lamb waves, which differentiates between symmetrical and anti-symmetrical modes (Fig. 7.12), the resolution of which shows several solutions according to a frequency f.

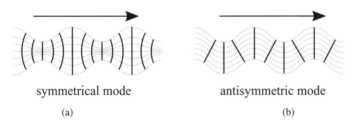

symmetrical mode   antisymmetric mode

(a)   (b)

Fig. 7.12 Symmetrical (a) and antisymmetric (b) Lamb waves

The results can be presented as curves corresponding to the phase velocity of
Lamb waves according to the product frequency-thickness (Fig. 7.13). This graph is
useful for several frequencies and plate thicknesses.

Thus, depending on the wave frequency and plate thickness, we obtain several
potential modes, that is, Lamb waves with different propagation velocities. For example, working with a 1 MHz frequency on a 2 mm steel thickness, that is, the product
$f \times e$ of 2 MHz·mm, can generate three possible modes: two antisymmetric A0, A1,
and one symmetrical S0. When the waves are generated in water, Snell's law (cf.
§ 7.1.4) gives the incidence angles for which these modes are obtained (A0 at 33.6°,
S0 at 17.8°, and A1 at 9.2°), considering a transmitted angle equal to $\pi/2$.

Generally, the easiest modes to decouple (they also are very energetic) are the
A0 and S0 modes. But, the simultaneous presence of several modes often leads to the
juxtaposition of waves, making the analysis (or the processing) of the signal more
difficult and the use of simulation necessary.

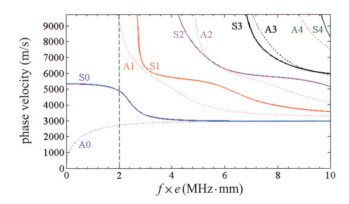

Fig. 7.13 Phase velocities of Lamb waves in the case of 316L stainless steel: example of propagating
modes for a product $(f \times e)$ that is equal to 2 according to the incidence

7.2   TECHNOLOGY AND EQUIPMENT

7.2.1   General information about ultrasound generation and part coupling

The generation of ultrasound consists in creating displacements of matter and reception corresponding to the recording of these displacements. The emission can be created with either a mechanical (ball, hammer, etc.) or thermomechanical shock (laser, electromagnetic micro-waves, etc.). However, these means are not often used because they generate spherical waves with amplitudes, frequencies, and directions that are difficult to control.

Another solution consists in connecting the part, via a coupling layer, or *couplant* (water, glue, oil, or grease), with a vibrating element such as a piezoelectric transducer. The latter turns the electric energy into mechanical energy and conversely. This principle is most often used in ultrasonic testing.

Multi-element probes (phased-array probes) composed of an active surface cut into many (piezoelectric) elements are often used, which allows the building of particular beams (focusing, deflection) by the judicious shift (suitable time-delay law) of the shots of each element. There are linear, annular, or 2D matrix probes, with elements becoming smaller and smaller. A low "pitch" (the distance between elements) generates beams in all directions but controlling such an important number of elements then becomes necessary.

Manual coupling is never constant. One solution consists in immersing both the part and the transducer, but this only applies to parts able to stand immersion. For larger parts, partial immersion is used with the ultrasound beam carried by water jets.

To find a solution for the coupling issue, methods of non-contact ultrasound have been developed, for example, *air coupling*, EMAT, or lasers. There are non-contact sensors on the market that use air as a coupling agent. New active materials with judicious impedance matches have been developed in order to reduce the acoustic contrasts between the transducer and air, and thus, the transmission of ultrasound to the inspected material, where the same contrast issue occurs. This currently limits the test to relatively low frequencies (50 kHz – 5 MHz) and to low acoustic impedance materials (composite, wood, concrete).

In conductive parts, eddy currents can be generated by *EMAT* transducers (electromagnetic acoustic transducer), and the interaction with the magnetic field of their electromagnets is able to create sufficiently energetic ultrasonic waves.

The generation of ultrasound by *laser* consists of a thermo-acoustic conversion (Fig. 7.14): a laser flash lights the surface of the medium and warms it, which rapidly expands the matter locally (thermoelastic domain), thus generating an ultrasonic wave in the propagation medium itself. Be careful not to enter into the ablation domain, as the test will no longer be "non-destructive"!

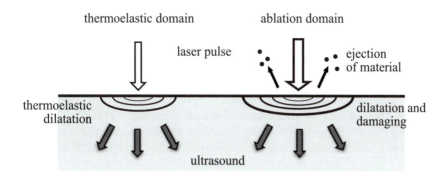

Fig. 7.14 The two domains of laser-material interaction

7.2.2 General information about the reception of ultrasound

The issue in reception is not about setting the matter into movement but measuring the displacements of zones on a given surface. This measure can be obtained in the same way as for emission: optically (lasers), mechanically (piezoelectricity), through air, or with electromagnetics (***EMAT***).

Apart from laser technology, which needs interferometry on reception, the great advantage of the ultrasonic testing method is that it needs only one emitter/receiver transducer (in ***echo mode***), which means that only one access to the part is necessary, even when the part is large.

Sometimes, the separation of the functions and the use of two transducers (***transmission mode***), with an emitter and a receiver using the same technology, is required. This is the case, for example, for the inspection of thick materials with high attenuation for which the double path (round trip) of ultrasound is impossible in terms of measurable energy.

7.2.3 Piezoelectric transducers

This is the most commonly used technology for ultrasound generation and reception. ***Piezoelectricity*** refers to the appearance of electric polarization at the surface of crystals that are subjected to a deformation (direct effect). Conversely, a deformation is obtained after polarization. Piezoelectricity appears in crystals that do not have a center of symmetry: compression or shearing separates the centers of gravity from the + and – particles which causes, at a macroscopic level, the polarization of the crystal surface.

The crystal cut (X or Y cuts) enables pressure waves (P) or shear waves (S) to be obtained directly at the output of the transducer. A mixed cut allows "P+S" transducers to be obtained.

The piezoelectric effect produced by natural materials (quartz, tourmaline, etc.) is very small. This has led to the development of ceramics with improved properties, such as barium titanate or lead zirconate titanate (LZT). Polyvinylidene fluoride (PVDF), a thermoplastic polymer with interesting piezoelectric properties, is flexible and can be manufactured in large quantities at a moderate cost.

The piezoelectric effects only occur below the ***Curie temperature*** (Tc); above this temperature, a phase transition occurs. The material goes from a ferroelectric state to a paraelectric state and the piezoelectric effect disappears. This temperature (often low, ranging from 120 to 300°C) is one of the main criteria in the choice of transducers, notably for measurements in hot environments. Note that the French Alternative Energies and Atomic Energy Commission (CEA – Commissariat à l'énergie atomique) has developed a high temperature ultrasonic transducer (*TUSHT – Traducteur Ultrasonore Haute Température*) using a lithium niobate crystal (Curie temperature 1140°C) in a sealed container, enabling direct functioning in liquid sodium at 550°C.

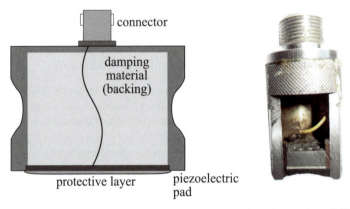

Fig. 7.15 Principle of piezoelectric transducer with cutaway view of a transducer (LCND)

The ***protective layer*** (Fig. 7.15) mechanically protects the active piezoelectric element. It also helps the ultrasonic transmission via an accurate calculation of the acoustic impedance (§ 7.1.4). The mass at the back is called a ***damper*** (or backing) and has two major objectives:

- It changes the emitted vibration to obtain pulse or sinusoidal rates. High damping improves the axial resolution in order to locate defects or measure thickness, for example. Low damping, on the contrary, improves the sensitivity of detection to measure attenuation, for example, to characterize a material or damage.
- It stops vibration right after the emission, so that the sensor is immediately able to receive ultrasound in return. High damping enables a low dead zone, that is, a small non-inspected thickness of a part under the transducer. Defects close to the surface can then be detected.

7.2.4   Generation of an ultrasonic field by a circular transducer, diffraction

In physics, when the wavelength gets close to or exceeds the dimension of the emitters (or the objects hit by a wave), the hypothesis of geometrical optics can no longer be used. ***Huygens*** (and then Fresnel) showed that "any point of the wave surface of a

primary source can be considered as a secondary source emitting a spherical wave."
The waves coming from different secondary sources interact with themselves and
have an impact on the shape of the wave: this is the principle of ***diffraction***.

Diffraction characteristics therefore disrupt the propagation of waves, including
ultrasonic waves. This theory can only be easily discussed within the context of sinusoidal mode and for a circular emitter in a perfect fluid, because in the case of solids,
even those that are isotropic, the phenomenon is more complex due to transverse waves.

In a fluid, in sinusoidal mode

For a plane circular disc emitting in a fluid (notion of piston effect), each point of the
surface creates a spherical wave. These waves quickly phase, combining or destroying
themselves and constitute two characteristic fields: one is disrupted while close to the
piston and is called the ***near field*** (or Fresnel zone), and the other, more regular is called
the ***far field*** (or Fraunhofer zone). A one-off measurement of acoustic pressure via a
hydrophone, for example, at the level of the plane disc (an immersion transducer, for
example) of diameter D, illustrates the general behavior of the propagation (Fig. 7.16):

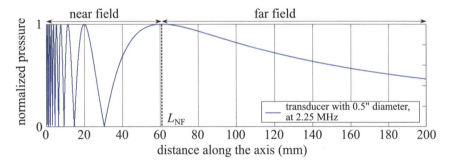

Fig. 7.16 Theoretical changes in acoustic pressure on the axis for a given frequency (sinusoidal mode)

On the axis of the beam, alternating pressure maxima and minima can be observed until the last maximum defining the ***near field length*** L_{NF}, from which the pressure decreases in an inverse function of the distance:

$$L_{NF} = {D^2}/{4\lambda}.$$

In ultrasonic testing, we try to work at this distance when possible. Outside the
axis, the transversal profile of the beam can be represented in the far field of a disc of
radius r and diameter D. For a given distance x to the transducer, the following ratio
is used: amplitude of pressure at an angle α to the amplitude of pressure on the axis
(Fig. 7.17). Thus, we obtain the ***directivity function*** of the emitter, according to the
wave number $k = 2\pi/\lambda$ with J_1 as a Bessel function of order 1:

$$\frac{P_{x,a}}{P_{x,0}} = \frac{2J_1(k\,r\,\sin\,a)}{k\,r\,\sin\,a}.$$

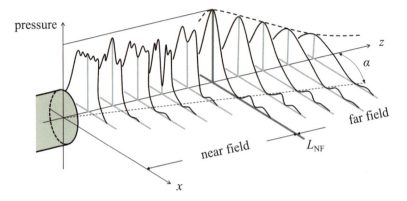

Fig. 7.17 Illustration of the acoustic beam of a plane circular transducer (according to French standards)

The acoustic pressure outside the axis is canceled several times. The first cancellation occurs for the first value that cancels the Bessel function of order 1, that is 3.83. The particular angle representing the **beam divergence** can be deduced via the semi-angle of aperture in the far field:

$$\sin \alpha = 1{,}22 \, ^\lambda/_D.$$

After this first angle, the notion of **side lobes** is generally introduced.

In an (isotropic) solid in sinusoidal mode
For a plane circular disc in a fluid (which is quite different for a solid), movements without deformation are considered. In the hypothesis that the pressure is uniform on the whole surface, we deduce that the shear zones are identical to the compression zones. In reality, this is not the case; unsteadiness occurs as well as the creation of side lobes (Fig. 7.18), which add to the main lobes (and to the side lobes mentioned in the previous paragraph).

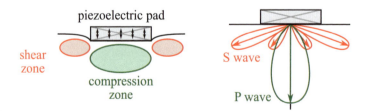

Fig. 7.18 Diagram representing directivity with main and side lobes

Pulse mode
In this case, characterized by the presence of multiple frequencies (broad band), the waves emitted by each point of the transducer can cross a point M on the axis, but

not necessarily at the same time, which creates a complex interference. To simplify the representation of the emitted field, the juxtaposition of individual propagations obtained for each frequency can be taken into account.

The profiles are then overlapped and the curve is smoothed. The near field is not disrupted as much, and in this case, there are fewer risks of misinterpretation of the test results.

7.2.5 Visualization of ultrasonic information: A, B, and CScan

Ultrasonic generator and receiver

Systems of emission are adapted to the type of transducers used and to the required methods of inspection. The generators can be independent boxes or emission cards inserted into a computer system or integrated into devices dedicated to ultrasound testing.

In the most common cases, the emitted signal is a pulse controlled in amplitude and in damping. The operating sensor emits a signal at the frequencies centered on the resonance frequency of the piezoelectric quartz disk. Frequency piloting is less common in industrial testing, but it is possible thanks to the generation of sine waves, possibly damped. Other types of signals can be created (Fig. 7.19), such as *chirps*, which combine frequency variations and even amplitude variations of a sine wave according to time.

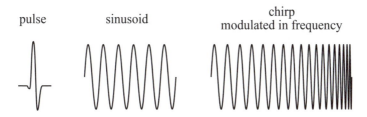

Fig. 7.19 Example of emission signals

Visualization of the AScan signal

A simple oscilloscope collects and displays the received signals. By definition, we talk about *AScan* (or A-mode, representation in amplitude) for signals for which the evolution of amplitude according to time is measured (Figs 7.19, 7.20, and 7.21). This electric amplitude is directly proportional to the amplitudes of displacement of the ultrasonic wave.

The vibrating nature of ultrasound waves gives oscillating signals presenting positive and negative parts. We talk about HF representation (for high frequencies) of AScan. In standard industrial testing where phase information is not necessary, a rectified and filtered AScan signal (called RF) is used. It is easily interpreted as it takes advantage of the available screen surface. On the other hand, frequency information is lost.

Visualization of BScan and CScan images

The *BScan* image is also called B-mode (brightness), according to the principle used by the first medical echographers. The scanning of a part is made according to one direction, either mechanically with a mono-element transducer, or electronically with a multi-elements (array) transducer. The image (space, time) is then created by juxta-posing the signals side by side and by assigning a color to each level of amplitude. The AScans are the columns of the image that show a sectional view (Figs 7.20 and 7.21).

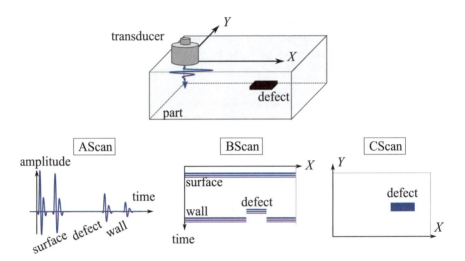

Fig. 7.20 AScan, BScan, and CScan representation of the same part (immersion)

A *CScan* image, C for cartography, is carried out with an XY scanning of the part. An image (space, space) is then created from the recorded value of each measured signal and converted into a level of color. Generally, the maximal amplitude in a given time window is used. This cartography globally corresponds to the projection on a plane of ultrasonic information obtained for the thickness of a part (Figs 7.20 and 7.21).

By definition, a "classic" industrial inspection is carried out on manufactured parts of near perfect ELHI materials that are globally homogeneous. Ultrasonic cele-rity is then considered constant according to the place of inspection and/or during the propagation. In this case, the "time" parameter of AScan and BScan representations can be converted into travelled distance, such as, for example, the depth reached in the part. But multiple echoes need special attention as they are difficult to handle.

7.2.6 The different types of transducer

These are called *transducers* because generally there is a transformation of energy (for example, electric energy turned into mechanical energy, for piezoelectric trans-ducers). But they are also commonly called *probes* or sometimes *sensors*. Different types of transducers are shown in Figure 7.22.

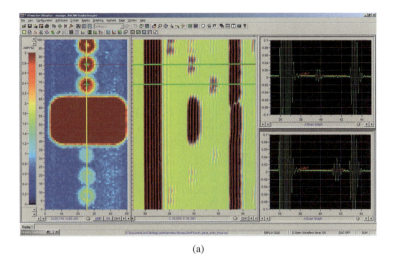

(a)

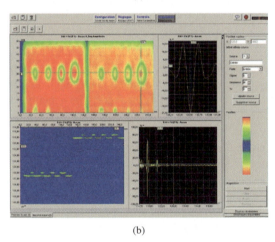

(b)

Fig. 7.21 Representations of AScan, BScan, and CScan. (a) The 2 AScans correspond to the 2 green crosses on the CScan, and to the 2 green lines on the BScan. The BScan comes from the section line (yellow) on the CScan. (b) Other system of multi-representations (M2M).

The main characteristic of a transducer is its frequency, which is the resonance frequency of the piezoelectric blade, its thickness being equal to $c/2f$, with f the frequency and c the speed of sound in the blade. Thus, high frequency transducers are particularly fragile. The choice of the work frequency results from a compromise: the highest possible frequency is chosen (thus improving the sensitivity of detection) compatible with the necessary propagation (including attenuation), which tends to favor low frequencies.

Then, depending on the purpose of the test, the geometrical parameters such as the angle of incidence and the diameter will be selected. The angle of incidence is chosen in order to perpendicularly approach the defect (maximal reflection), and the diameter is chosen according to the dimension of the part, the depth of the test (near field length), and the dimension of the detectable minimal defect.

Fig. 7.22 Different probes: straight-beam, fixed angle-beam, interchangeable wedges, immersion, phased-array; a hydrophone; a cutaway transducer (LCND)

A ***straight-beam probe*** generates waves that propagate perpendicularly to its active face. Positioned on the part with contact, via a couplant, it is used for delamination-type defects, which are parallel to the faces. The generated waves can be longitudinal (most of the time) or transverse (specific testing) according to the cut plane of piezoelectric crystals.

Damping is chosen according to what is needed: it is low when sensitivity needs to be increased or high when improving the resolution in depth is what counts (§ 7.2.3). Sometimes, an acoustic relay can be found bonded to the active face in order to shift the emission (and then the ***dead zone***) and to detect the defects close to the surface. If the dead zone is to be further reduced, the transmitter will not be used as a receiver (and it keeps vibrating). Instead, sensors combining separated and slightly shifted emission and reception will be used.

For inclined defects, or when the geometry of the part does not allow perpendicular access to the defect (welding bead, for example), an ***angle-beam probe*** is used. The latter is usually employed with the help of a normal probe that is made to transmit in a plastic guide (Plexiglas, etc.) machined to a specific angle so that the desired waves are obtained when changing mode at the inspected plastic-material interface.

Inspection beyond the first critical angle (§ 7.1.4) means that there is only one wave (S) to handle and therefore less parasite echoes. However, although it is sometimes better to favor the less attenuated P waves, the S waves will still be present. Straight-beam probes with fixed angle (S waves in general), with attached or interchangeable wedges or variable incidence probes (P +/– S) and with a multi-position wedge guide system are available.

For ***immersion testing***, adapted P probes are used (not S because they do not propagate). The main differences concern the acoustic adaptation blade that provides ideal transmission in the water and in the connections in the waterproof box system. An integrated lens can be added on the front face in order to concentrate the ultrasonic energy.

A piezoelectric quartz plate can also be shaped (but this is less common). Thanks to the focusing, the *focal zone* of the *focusing probe* is then very small (Fig. 7.23); its length and diameter are defined by the **length at –6dB** and the **diameter at –6dB**:

$$d_{-6dB} = \lambda F_{ac}/D$$
$$L_{-6dB} = 4\lambda(F_{ac}/D)^2 \text{ if we consider a cylindrical zone.}$$

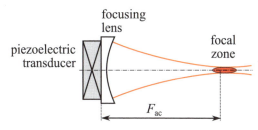

Fig. 7.23 Acoustic focusing

Because of the real phasing and oppositions, the acoustic phenomena are similar but also different from the optical phenomena. The **acoustic focal length** is notably different from the **optical focal length**, obtained by following the light rays through the lens. Thus, a normal probe, that is, a non-focused one (optically), can also be considered as focusing (acoustically) in the near field length (L_{NF}), with a useful beam diameter given by the relation $d_{-6dB} = D/4$. The displacement of energy to the transducer (the focusing) can be expressed by the following *focusing* $g = F_{ac}/L_{NF}$.

In order to adapt to complex geometries of parts and to all angles of incidence, bi-focusing even tri-focusing lenses can be manufactured to reach phasing of a slightly distorted focal zone (cylindrical) at the desired location on the part.

A **multi-element** or **phased-array** probe is a collection of small active transducers, the number of which complies with a 2^n law (32, 64, 128, and so on) related to digital control. The design of these sensors (Fig. 7.24) is diverse: linear, matrix, concave, etc.

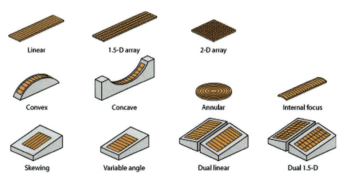

Fig. 7.24 Different types of multi-elements ultrasonic probes (OLYMPUS)

Control, in emission and reception, is carried out by adding adapted delay laws. Each element is the source of a spherical wave, and the assembly, controlled in consequence, makes scanning, focusing, and/or deflection of the beam possible.

7.2.7   Impulse response and transfer function – bandwidth

It is useful to know the *impulse response* of the ultrasonic transducers used, notably for checking their conformity with simulations or their possible drifts. The response has to be checked prior to each important step of a test. Knowing the response can help in the decision to replace, if needed, the transducers of an automatic installation.

The impulse response is the reaction of the whole system (generator, transducer, wires, receiver, and visualization system) to an impulse (Fig. 7.25). In industrial inspection, only the AScan obtained for a classic impulse from the generator, on a ball (immersion), a side drilled hole, or a block backwall (with contact) is checked.

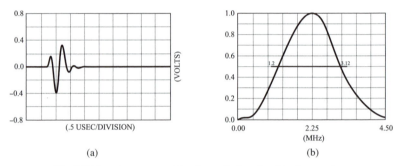

(a)   (b)

Fig. 7.25 Impulse response (a) and transfer function (b) of a transducer

The *transfer function* is the frequency response that is obtained by a Fourier transform of the impulse response. The (amplitude) spectrum obtained has a certain width. Halfway up, for example, (that is, an amplitude corresponding to that of the maximum, reduced by 6 decibels), we obtain two frequencies that define *–6dB bandwidth*.

A transducer with a rather low damped impulse response (which looks like a sinusoid) will have a frequency response similar to that of a sinusoid, that is, a unique value (in signal mathematics this is a Dirac). This is called a *narrowband*.

Conversely (the formula $1/T$, which refers to the fact that frequency is the reverse of time, makes sense!), a transducer having a rather high damped impulse response (which looks like a Dirac) will have an almost infinite frequency response. This is called *broadband* (Fig. 7.26).

We have seen how a transducer with a "high damped" impulse response was useful for increasing resolution in depth (i.e., temporal resolution). It can now be added that as it is in "broadband," the beam is composed of several frequencies and that each one of them will follow its own speed and attenuation laws, etc. and finally, that the comparison with the theory (always validated in sinusoidal regime) will be more complex.

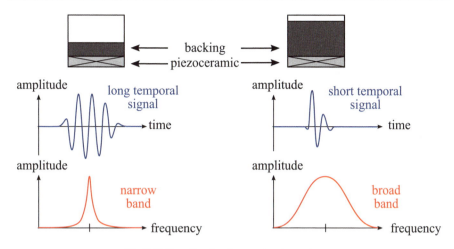

Fig. 7.26 Broadband and narrowband transducers

7.3   METHODS OF TESTING

7.3.1   Calibration and ultrasound

As already explained in paragraph § 2.4, the qualification of an NDT method is obtained by comparing the results to those obtained from a standard part which contains standard defects. The set has to be representative and manufactured at low cost. In industrial ultrasound testing, the detection of defects is based on the laws of (specular) reflection on input interfaces of the defect. So two requirements have to be met:

- The *standard part* must have the same dimensions and same material as those of the real part, so that the evolutions of propagation are known: attenuation, structural noise, etc. A homothetic transformation (a smaller standard part) is considered only if the material is perfect (ELHI), or considered as such. Otherwise, the reproduction at a smaller scale will not be accurate because the evolution of propagation (cf. chapter 2) is essentially linked to the metallic characteristics of the material or structure (welding, for example), which are themselves closely related to mass and dimension.

- The *standard defect* must have the same shape as a real defect and must be of a minimal dimension. But as ultrasonic detection is based on wave reflection at the level of the first encountered interface (beyond this the presence of air generally forbids any propagation), ultrasound only needs a wall (which is easy to create) to simulate a crack or a delamination (Fig. 7.27). For ultrasound, a wide machined groove may perfectly simulate a real delamination, which is very hard to manufacture.

Thus, in industrial inspection, different simple *reference blocks* (Fig. 7.28) are available on the market. These are used to check the material (linearity), the transducer (directivity, attenuation), and to determine the sensitivities of detection (*side drilled hole*, *flat bottom hole*, notch). All these methods are governed by international standards.

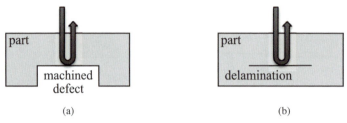

(a)   (b)

Fig. 7.27 A manufactured notch (a) representative, for the ultrasound, of a real delamination (b)

Fig. 7.28 Various reference blocks (LCND)

In expertise, for the sizing of a crack, for example, using the precise location of the echoes of diffraction of the edges (§ 7.3.5), ultrasound testing again presents the advantage of only requiring a standard, easy-to-manufacture defect. The diffraction can be studied on an electro-eroded notch, easy to manufacture because it diffracts less than a real crack, which is more difficult to manufacture. As this type of calibration is pessimistic, it can be validated.

As a conclusion, ultrasound testing generally settles for the use of simpler reference defects than other NDT methods, that is, defects that are easier to manufacture. It is another reason to consider ultrasound testing as the most efficient method for solving today's complex NDT issues.

7.3.2   Contact techniques: echo, transmission, tandem

Contact techniques are fast and easy to implement. Equipment is light and well adapted to manual testing. These are the most commonly used techniques despite the problems caused by coupling, which makes the precise evaluation of the amplitude of detected echoes difficult, even when trying to make coupling as constant as possible.

On contact, in a manual testing, ***pulse-echo*** techniques are generally used where the transducer is both transmitter and receiver. It is not necessary to have access to the two faces of the part, as has previously been mentioned. It is one of the best advantages of the ultrasonic method. Normal or angle probes can be used.

Handling two transducers, where one is transmitter and the other receiver ("pitch-catch" technique), is more difficult (Fig. 7.29). In this case, the advantage is in the fact that the ultrasonic wave has only a one-way path to follow, which helps test some highly attenuating materials.

The ***through-transmission*** technique is also used with two angle transducers located on the same face of the part. This ***tandem*** assembly is often used, notably in TOFD techniques (§ 7.3.5). In these pitch-catch cases of through-transmission (or tandem), the logic of the inspection is changed; this time there is a loss of signal because of the presence of defects, which is another way to automatically record the inspection that has been carried out.

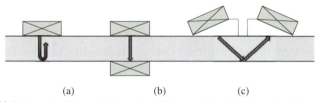

(a)   (b)   (c)

Fig. 7.29 Pulse-echo pitch-catch (a) or through-transmission (b) and tandem (c) techniques

7.3.3   Immersion techniques

The same principles as in contact techniques are applied here, but this time the part and the transducer are completely immersed in a tank containing the coupling liquid. If there is no major problem (corrosion, etc.) this liquid is water, preferably degassed water. Thanks to mechanical systems (Fig. 7.30), the transducer can carry out different automated or non-automated scans around the part.

Fig. 7.30 Ultrasonic immersion testing tank (LCND)

In comparison to contact testing, the major benefits of the immersion techniques are the possibility of using focused transducers and of mapping the part (BScan, CScan) or some specific places of the part, in a more reliable and repeatable way (as opposed to manual testing).

The focusing of ultrasound, compared to the standard ultrasonic method, enables:

- The concentration of ultrasonic energy in a limited field of the part
- The reproducibility of the testing, as the focal zone is considered homogeneous
- The improvement of the measurement resolution as the focal zone can be very small

A reflector will be efficiently seen if it is located in the focal zone but the immediate inconvenience is that the time of testing is longer. For the inspection of thick parts, many passages may have to be made or the use of transducers with different focus zones should be used.

All types of testing can be done by immersion with or without focusing because the influence of the water-part interface can be calculated (Fig. 7.31a), which acts like a diopter: normal and oblique incidence, surface waves, Lamb waves, etc.

The calculation of the focal zones consists in obtaining a cylindrical focal zone in the part, that is, of identical diameters and lengths in the incident plane and in the perpendicular plane. To do so, a fictitious transducer has to be defined. It produces an acoustic field in the material similar to the field created by the real transducer (Fig. 7.31b).This fictitious transducer operates in a medium the impedance of which is equal to that of the part. The focal zones and optical dimensions (theoretic) are then calculated before being converted into acoustic dimensions (corresponding to reality).

To study and characterize the ultrasonic beam, focusing or not, the evolution of amplitude in the beam has to be known. To do so in immersion, it is possible to measure for each point of the emitted field, the amplitude, the time of flight, etc. using a hydrophone. This type of transducer only works as a receiver. Sometimes it is simply replaced by a reflection made on a small size ball (defined by a standard). However, note that a plane, as it reflects the entire information from the beam with an averaging of the transverse amplitudes, does not really help in knowing the ultrasonic amplitude at a specific point of the axis.

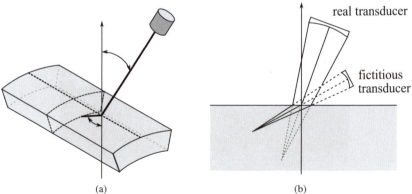

Fig. 7.31 (a) External testing of a cylinder: with an incident plane containing a generatrix (b) calculation principle: real and fictitious transducer

7.3.4   Multi-element technology

This technique brings solutions to numerous testing issues. It accelerates the rate as different zones of a part can be insonified without mechanical movements of the transducer. The coupling is then constant.

Scanning and deflections (Fig. 7.32) can be done in one or two planes using linear array probes or multi-element matrix probes. Focusing (electronic) is possible in contact or immersion.

Recently, flexible multi-element (conformable) transducers have been developed by the French Atomic Energy and Alternative Energies Commission (CEA). They ensure a controlled propagation (automatic adaptation of the delay laws), despite the evolution of the geometry of the part's surface (Fig. 7.33).

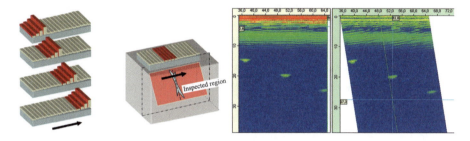

Fig. 7.32 Laws of delay and electronic linear scanning, and BScans obtained in normal and oblique incidences (M2M)

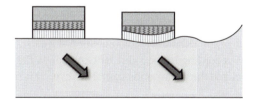

Fig. 7.33 Principle of conformable transducers

7.3.5   Analysis of ultrasonic information

In the first chapter, we saw that generally, the operation of NDT (at least that of defectometry) is carried out in different steps according to the needs. First the (indispensable) detection of elements that are potential defects, and if needed, after an analysis, the location, the identification of their shape (volume or non-volume), and the sizing of the defects to deduce their noxiousness.

Knowing the propagation velocity (the absolute value or according to a standard part) and the direction of the propagation allows the location of the defects in depth.

Further traditional techniques, mainly based on the exploitation of the amplitude of the ultrasonic signal, determine the shapes and dimensions of the reflectors or at least, the reflecting surface (that is, the surface perpendicular to propagation).

Geometric characterization of defects

Characterizing the geometry of a defect can consist in rotating the transducer around the reflector (orbital technique) starting from the maximal amplitude position. Amplitude drops very quickly with non-volume defects (for example, a plane crack) as opposed to volume defects such as inclusions or blowholes.

The sizing of big (plane) defects: –6dB method

This sizing takes into account the amplitudes pointed out when the beam is reflected by the defect. The *–6dB method* consists in plotting the field reflected by the defect: it is maximal when the whole beam is reflected and divided by 2 (i.e., reduced by 6dB) when half the beam is on the defect and the other half is outside (Fig. 7.34a).

With this method, the sizing of a predominantly plane defect is possible only if its reflecting size is bigger than the size of the beam width. In accordance with paragraph 7.2.6, for small defects, the size of the beam has to be reduced and a priori the transducer's diameter ($d_{-6dB} = D/4$), which increases divergence. The use of focused transducers is favored, the small diameter of their focal zone (a few 1/10 mm) can be obtained, for example, with an increase in the diameter of the transducer.

The sizing of small defects: AVG or DGS method

When a defect is smaller than the beam, the AVG method (*Abstand-Verstärkung-Größe*) or DGS (distance-gain-size) can be used. It compares the amplitude reflected at the backwall of the part in the presence of a defect with the amplitude reflected when there is no defect. This comparison made via an AVG diagram, defines for the defect a diameter of an equivalent circular reflector, giving the same background echo attenuation. In this diagram (Fig. 7.34b), attenuation A, in dB, at a given depth D, in millimeters, changes according to the value of G, which is the ratio of the estimated diameter of the defect and the ultrasound beam. However, this method underestimates the real size of the defect.

The precise sizing of plane defects: diffraction echoes

If a plane defect (crack, delamination) is insonified with a non-null incidence, echoes of diffraction appear from the tips of the defect. If the crack is an emerging one (Figs 7.35 and 7.37), the *corner echo* that locates the output of the crack has to be found in the first place and the *diffraction echo* of the crack tip. The distance between these two positions is linked to the size of the defect.

This method is accurate, but the amplitude of diffraction echoes is lower than that of specular reflections or other corner echoes, which makes them difficult to detect or even undetectable if the signal-to-noise ratio is too low. Note that diffraction echoes at the top and bottom edges of a crack, being in an opposition of phase, are different from the echoes of two blowholes, for example (same phase), which are less noxious defects.

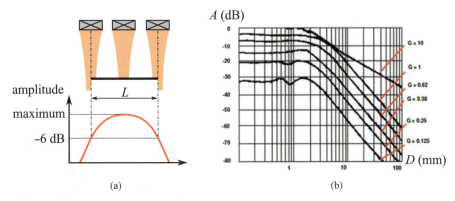

Fig. 7.34 "–6dB methods" for plane defects (a) and "AVG method" for small defects (b)

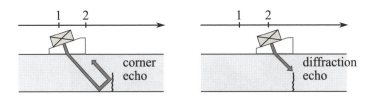

Fig. 7.35 Sizing of an emerging crack

Analysis of BScan and CScan images

The easiest image to interpret is a CScan image. It is a 2D image (space, space). It represents an "ultrasonic top view" of the part with information corresponding to the amplitudes reflected according to a certain incident angle, for a part thickness and converted into colors. So, there is definitely "equivalence" between the visible form of the image and the real form of the defect, but we should bear in mind that the ultrasound reflection can follow different and complex paths. Figure 7.36 below shows a delamination (yellow and red) in a composite part.

A BScan is more difficult to interpret as it is a 2D space-time image. The conversion of the time axis into space axis (depth, for example) is possible only if the celerity of ultrasound and its path are known. With those precautions, a BScan image generally represents an "ultrasonic sectional view" of the part.

Figure 7.37 shows a BScan image of an immersion test under an oblique incidence (45° in the part). The part, T-shaped, has a groove (C position), that is clearly detected. The BScan here hardly shows (as opposed to the AScan where it is sharp) the presence of the calibration crack (electroeroded, E position) and of different echoes caused by reflections on the shape of the part. Generally, the BScan is sufficient in itself to archive the inspection but its interpretation for a diagnosis can be difficult. The AScan, at the origin of the image, is favored for measurement and comparisons.

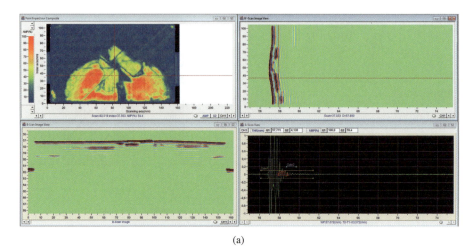

(a)

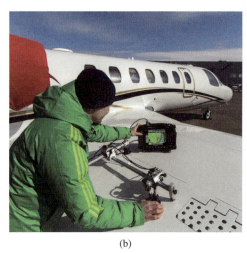

(b)

Fig. 7.36 CScan of a composite and detection of delamination (a) (MISTRAS), example of use in an industrial context (b) (OLYMPUS)

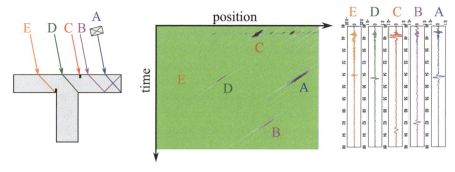

Fig. 7.37 T-shaped part and associated BScan and AScan

TOFD method

The TOFD (time-of-flight diffraction) method detects defects like cracks via the diffraction echoes obtained from their tips. When the two transducers coupled in tandem (pitch-catch) pass over the top of the defect, each tip of the crack produces an ellipsoidal signal in the BScan image (Fig. 7.38). The top of the ellipse (which means the shortest transit time) is obtained when the tandem is at a position symmetrical to the crack.

This method is efficient and easy to use because diffraction echoes are less sensitive to the incident angle on the defect than direct reflection echoes. But this technique has some limits, especially for emerging defects or defects close to the access surface, because the difference in time of flight between the waves is too small.

Reading of the image is simple, and many people consider the TOFD method to be a substitute for radiography, notably for this type of defect orientation, which is typical of damage caused by fatigue.

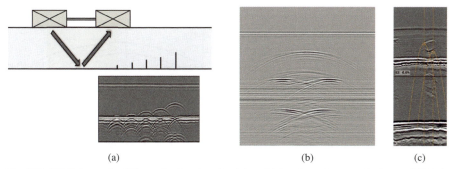

(a)   (b)   (c)

Fig. 7.38 TOFD image for different depths of notches (a); multiple echoes due to mode changes P-P, P-S, S-S, etc. on an electroeroded notch (b) (MISTRAS); use of an adapted cursor to identify the ellipses (c) (SOFRANEL)

Need for the digital processing of information (signal, image)

The tools of analysis (for detection, characterization, etc.) introduced in this paragraph, whether for AScan, BScan, or CScan, are simply based on the consideration of the amplitude of temporal signals. But there are cases where this is not enough, for example:

- When the amplitude of the defect signal is low according to the amplitude of the events around the defect. This is a low *signal-to-noise ratio* (**SNR**). There are several definitions for SNR but the simplest refers to the ratio between the amplitude of the defect's echoes and the maximal amplitude of the noise's echoes. Some codes consider that the ratio has to be higher than 2 (or even 3) so that the operator can make the decision without hesitation, regarding the presence of a defect.
- When the problem is very specific and cannot be expressed by the "shape" of the AScan only. Searching for the evolution of the elasticity module gradient of a thin nitrided coating, for example, requires the calculation of "specific" parameters in the signal, such as phase velocity by frequency, etc.

- When frequency parameters are dominant. For example, the measurement of attenuation of a material can be calculated from the evolution of ultrasonic amplitude during propagation, but the (different) result that takes the frequency amplitudes into account complies more with the physics of the phenomenon. Thus, for an austenitic stainless steel part (thickness 22 mm), inspected with a 2.25 MHz transducer, the attenuation (Fig. 7.39) is 175 dB/m if it is measured on back and forth echoes, or it is 99 dB/m for 1 MHz, 145 dB/m for 2 MHz, and of 250 dB/m for 3 MHz if it is calculated on amplitude spectra. On one hand, these latter values are now independent from the used transducer and on the other hand, they show results that are more in line with the reality of the attenuation phenomenon, which is closely linked to frequencies.
- When an operation is to be automated, for example, detection according to a certain threshold.

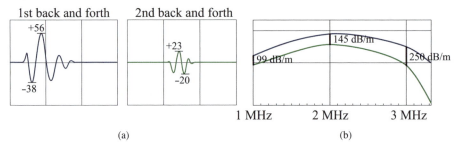

Fig. 7.39 Attenuation measures for temporal (a) or frequency responses (a)

In all these cases, digital processing (signal or image) can be considered in order to find solutions. The study of these different cases, like all other complex cases of the degradation of information, is explained in chapters 13 and 14. The main NDT methods (radiography, etc.) are all presented, but ultrasound testing is outlined in detail.

7.4 EXAMPLES OF INDUSTRIAL APPLICATIONS

The introduced applications show the diversity of ultrasound testing: there are classic industrial cases and more complex expertise cases. The perspectives of applications coming from research will be explained in paragraph 7.5.

7.4.1   Weld testing

Ultrasound testing is suitable for testing welds (Fig. 7.40). Most of the defects resulting from weld execution (cold and hot cracks, blowholes, inclusions) are detectable,
even if radiography (for inspection into thickness), penetrant testing, or magnetic testing (detection of emerging defects after each pass), are competing techniques.

Fig. 7.40 Example of weld testing (OLYMPUS)

For large size thickness (more than 10–20 mm) or when searching for the accurate dimensions of a crack, ultrasound testing is favored as opposed to radiography,
which is more dangerous. The ability of ultrasound testing to adapt to the orientation
of bonding defects is also appreciated.

7.4.2   Inspection of bars and tubes with phased array probes

Multi-element, concave, or fixed probes can replace the rotating heads that are sometimes used. They generate (Fig. 7.41) an ultrasonic scan, which, on demand, can be
in P0 incident wave (for the detection of volume defects) or S45 (for surface defects).

Fig. 7.41 Inspection of tubes via phased array probes (OLYMPUS)

7.4.3   TOFD inspection of weld

The TOFD method (time-of-flight diffraction, see § 7.3.5) replaces radiography when the pipes and pipelines are laid and installed (Fig. 7.42). It enables sizing defects in order to study their noxiousness.

(a)   (b)

Fig. 7.42 TOFD manual version (a) and automatic version (b) (MISTRAS)

7.4.4   Measure of thickness and corrosion

The ultrasonic *measure of thickness* usually requires using separated emission/reception transducers in order to make "pseudo-focusing" possible. Uncertainty is in the order of 0.1 mm in steel, but it can be smaller if the conditions related to the homogeneity of the material, residual stresses, temperature, and roughness are under control.

Thickness measurement meters (Fig. 7.43) measure the time of the path of a longitudinal wave (P) via an internal clock. The automatic display of a thickness takes into account the estimated ultrasonic velocity of the material, which is given via a test on a representative gauge block. Visualizing the ultrasonic signal is not necessary except for some cases such as measurements under coating.

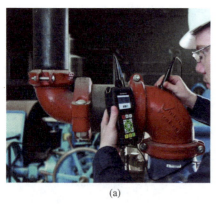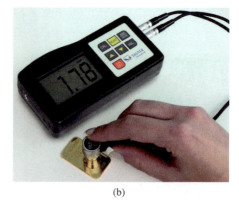

(a)   (b)

Fig. 7.43 Thickness measure with ultrasound (a) (OLYMPUS), measuring the thickness of a gold plate (b) (SAUTER)

Figure 7.43b shows a TD-GOLD meter which determines if gold bars or coins are authentic or fake, hiding metallic or non-metallic cores. This device uses ultrasound to measure the density of gold bars or coins.

7.4.5   In-service inspection machine (MIS in French)

INTERCONTROLE carries out all the inspections of the reactor vessels of the French nuclear power plants (this activity has been expanded to Daya Bay, China, and Belgium). INTERCONTROLE developed the *in-service inspection machine* (currently MIS 7 and MIS 8) to carry out an inspection from the inside of a vessel. These 12-ton robots (Fig. 7.44) are equipped with all the necessary tools for testing (radiography, ultrasound).

Welds and their thermally affected zones are tested by focused ultrasound to detect defects under coating. There are two types of software that analyze the ultrasonic data recorded during the inspection: Civacuve for all the tests made on a vessel and Civamis for expertise inspections.

Fig. 7.44 MIS and its ultrasonic equipment (AREVA)

7.4.6   Acoustic microscopy

Acoustic microscopy is an imaging method using high-frequency ultrasonic waves in the range of a few MHz to several GHz. There are two techniques: ultrasonic micro-echography and ultrasonic micro-interferometry.

Micro-echography is the ultrasonic testing of small dimension parts without any specific preparation such as polishing or chemical pickling. Generally used in reflection and impulse modes, the search (Fig. 7.45) is made on the surface or in depth. With frequencies from 500 MHz to 1000 MHz, the lateral resolution can be of a few micrometers and depth of a few dozen micrometers.

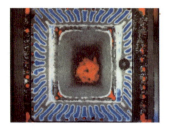

Fig. 7.45 Inspection of a semi-conductor, with delamination (in the center, in red) (C.GONDARD)

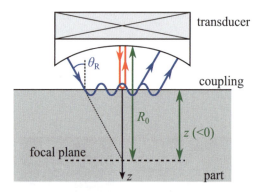

Fig. 7.46 $V(z)$ technique

Micro-interferometry gives access to the local mechanical characteristics of a material via the measure of the propagation velocity of Rayleigh waves. It is also called $V(z)$ technique, because the velocity is calculated from the measurement of the different arrival times of the surface waves, according to the setting of depth z (Fig. 7.46).

Its fields of application are numerous: microelectronics (multilayer components), metallurgy (damage), food-processing industry (roasting), and the biomedical field (biopsies in operating rooms).

7.4.7   Quick ultrasound testing with lasers (aeronautics)

There is no influence on the angles of incidence with the principle of laser generated ultrasound. This technique is therefore efficient for the inspection of complex or hardly accessible geometries (Fig. 7.47). The particularity of laser also makes it efficient in various hostile environments (high or low temperatures) and/or sometimes very distant parts (a few meters).

7.4.8   Inspection of the tightening of bolted assemblies

The ultrasonic evaluation of the tension of tightening depends on the measurements of the wave time of flight involved in the process. (Fig. 7.48). There are two simultaneous phenomena that increase the ultrasonic transit time for a tightened bolt: the screw

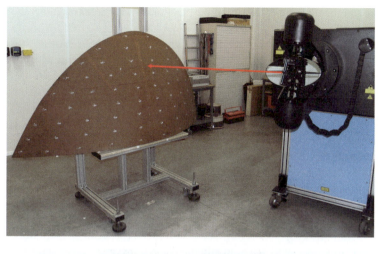

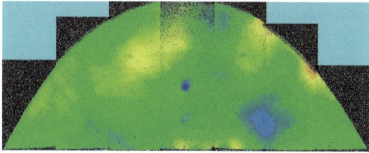

Fig. 7.47 Thickness map of a composite panel with double camber panel (LUIS system, AIRBUS)

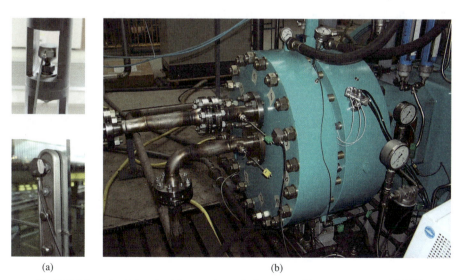

(a)   (b)

Fig. 7.48 Measuring tightening (a) on an air terminal structure, (b) on a compressor (CETIM)

is longer and the ultrasonic celerity of the steel materials decreases under the applied tension stress (acoustoelastic effect).

The difficulty comes from the lack of knowledge of the acoustoelastic characteristics of each screw, which change according to the material with which the screw is manufactured (end of a bar, full bar, machined or rolled threading, etc.). "Bolters" only have to comply with standards related to the mechanical resistance and not to the values linked to ultrasound.

So, either the characteristics are known and archived, and each bolt is identified before testing or they are not. In the first case, the ultrasonic method of evaluation of tightening is fast with low measure uncertainties (+/–3%). These values can be recorded (by marking) on the head of the screw, or the screw can be equipped with a piezoelectric element on its head.

In the second case, when there are no available acoustoelastic characteristics, carrying out a measurement without stress (by loosening the bolt) can provide the missing information. It is when loosening is not authorized (permanent operating machine) that LCND laboratory work using longitudinal and transverse waves showed that only a dual-wave method enables the evaluation of the stress with a reasonable uncertainty (which is all the lower as stress is high).

7.4.9   Inspection with "water jet" technique

For large parts that have to be quickly and entirely (100%) inspected, the "water jet" transmission technique is of particular interest. The part is positioned between two transducers, attached to synchronized robot arms, and the emitted wave is transmitted via an initial water spray.

Depending on the possibility for propagation in the part and the presence of a possible defect, part of the energy goes through the part and is transmitted to the receiver transducer through the second water spray (Fig. 7.49). A CScan-type map is created taking into account the transmitted ultrasonic energy according to the spatial position.

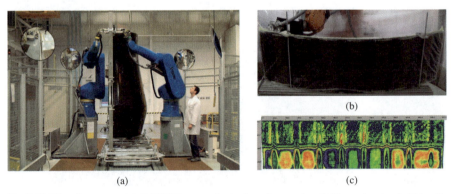

(a)   (c)

Fig. 7.49 Water jet system (a) (AREVA); water jet testing of a panel and CScan showing bubbles (in orange) on the carbon skin/honeycomb interface (b) (c) (SAFRAN-SNECMA)

The method is now adapted to parts with complex shapes thanks to the use of multiple hinge robots with edge tracking ability. It is essentially used in the aeronautic field on monolithic composite panels or on metallic or composite sandwich "honeycomb" structures. Lack of adhesive, and in the latter case, delamination in the skins are inspected.

7.4.10 Inspection with EMAT transducers

As already explained in § 7.2.1, ultrasound can be generated without contact in a conductive part. The advantage is obviously the reproducibility of the coupling and the potential speed (Fig. 7.50).

(a)   (b)

Fig. 7.50 EMAT inspection of a bar (a) and generation of Lamb waves (b) (SOFRANEL)

7.4.11 Guided wave inspection

The use of guided ultrasound waves in NDT enables the quick inspection and monitoring of structures over large dimensions in a global way. They are especially used for the inspection of metallic structures (tubes, pipes, wires, etc.), subjected to ageing due to mechanical fatigue and corrosion. Guided waves can propagate for a relatively long distance along the structure (then called wave guide), with the help of transducers positioned at one place in pulse-echo mode (Fig. 7.51).

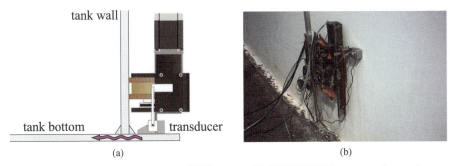

(a)   (b)

Fig. 7.51 TALRUT (a) (MISTRAS) and LORUS systems (b) (APPLUS RTD) to inspect the annular support of a storage tank

Zones with varying thicknesses (caused by corrosion, for example) or the presence of a defect (a crack) modify the propagation of guided waves and reflect a part of the energy that makes detection and location possible. The use of guided waves also means that accessible as well as less accessible structures (buried, insulated, coated, etc.) can be tested.

However, there are a few disadvantages to this method. Calibration is not always a simple operation. The interpretation of measurements can be difficult because of the distortion of signals in the case of complex geometries and the dispersive nature of guided waves. They can be highly attenuated by the structure environment (coating, etc.).

7.4.12 Synthetic focusing (SAFT, TFM, etc.) and adaptive methods

The total focusing method (*TFM*) is currently being integrated in a few phased-array technology probes. It is an extension of the idea of the *SAFT* method (synthetic aperture focusing technique), which consists in resetting the events found when scanning with a single-element transducer. If the events belong to the same defect, they finally phase according to the resetting curve, otherwise they are destroyed. As a consequence, lateral resolution is greatly increased, as is the case with real focus transducers.

The algorithm of TFM reconstruction is the post-processing (now carried out almost in real time) of a phased-array acquisition, usually of *FMC* type (full matrix capture), that is, when an element of the array is activated, all the others are "listening" in reception. The operation is repeated for the individual activation of each element of the linear array sensor. Every law of delay can be artificially a posteriori tested.

Compared to traditional imaging, the spatial resolution is improved and homogeneous at all points (Fig. 7.52). Adaptive TFM imaging is currently being developed and will make the testing of complex parts in immersion via conformable transducers possible.

The *SAUL* (surface adaptive ultrasound) method is the use of an iterative algorithm that was developed for the immersion testing of composite stiffeners. The technique is iterative: it consists in generating the first shot, measuring the times of flight between the elements of the transducer and the part surface, and in calculating the delay law that will be applied to the next shot, and so on, until the adaptation of the wave to the surface of the part is complete.

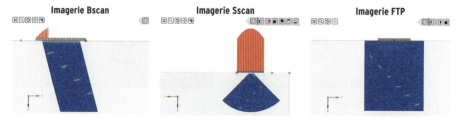

Fig. 7.52 Real-time TFM imaging; the SScan is a BScan with an angular scan (M2M)

Generally, the method enables the adaptation of any wave to any complex surface (Fig. 7.53) without requiring any precise information about the characteristics of the part and the probe. It remains valid for 2D (linear sensor) and 3D (matrix sensor) geometries.

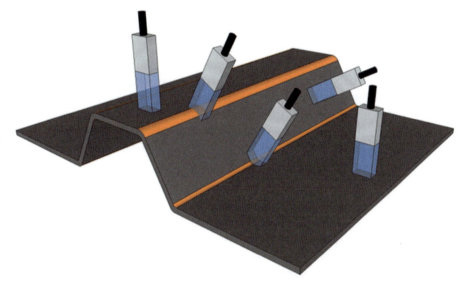

Fig. 7.53 Testing self-adapting to the geometry (M2M)

7.5   CURRENT AND FUTURE POTENTIAL

Industrial ultrasound testing, compared to the other NDT methods, offers several benefits such as:
- Quick and easy implementation, notably for contact techniques
- Access to the two faces of a part is not necessary
- The ability to go through very large thicknesses
- Good adaptation to the natural directions of most defects
- The 3D location of detected defects
- Accurate sizing of non-volume defects such as cracks
- Simplified creation of standard defects
- Imaging, not reserved to laboratory testing anymore

However, there are disadvantages. These include:
- The need to couple the transducer to the part
- The necessity for covering the whole part, as this testing is local and not global
- The high sensitivity of ultrasound propagation to heterogeneity and/or anisotropy levels of the material

The variety of this method and the diversity of the possible cases of applications, whether it is in NDT (detection of defects) or NDE (characterization of parameters), are the reasons why the use of ultrasonic information is often favored in order to solve the most difficult cases. The potential of ultrasound in measurements, characterization, and evaluation of physical parameters by far exceeds all the other non-destructive methods, notably because its sensitivity to the material is a major asset.

The analysis of ultrasonic data turns into a matter for experts because traditional methods taking temporal aspects into account (which are completely satisfying for classic cases) are unclear or unusable in these cases.

Chapter 13 focuses on the study of the factors responsible for the degradation of the measurements in NDT and/or NDE (including ultrasound) and that may distort the diagnosis. However, it also introduces suggested solutions that can be used. Chapter 14 will show an overview of solutions via modeling and digital processing.

The next paragraphs introduce two in-depth studies specific to ultrasound testing. They correspond to research carried out in the LCND laboratory and have a strong potential for future industrial applications.

7.5.1   Non-linear acoustics

The non-linearity of a material distorts the waves during propagation. In the pressure zones, waves can accelerate (according to the sign of the non-linear parameter) and in the tension zones, the waves slow down (Fig. 7.54a). All this makes the associated frequency representation change so that harmonics appear (Fig. 7.54b).

Figure 7.54c shows that non-linear indicators can be more sensitive to damage than the classic, linear indicators such as ultrasonic velocity or attenuation. The purpose of the study of the different non-linear parameters is to provide the earliest possible detection of damage in order to reduce the cost of maintenance.

The difference is made between "classic" non-linearity, resulting from non-linear interatomic force-distance relation, and "non-classic" non-linearity, caused by contact or friction at a microstructural scale and present in heterogeneous mediums such as rocks or concrete.

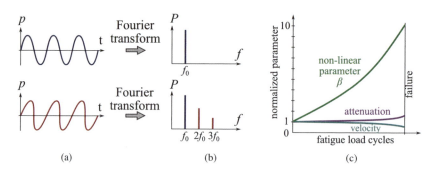

Fig. 7.54 (a) Distortion of a sinusoid, (b) creation of harmonics, (c) evolution of non-linearity, velocity, and attenuation when there is damage

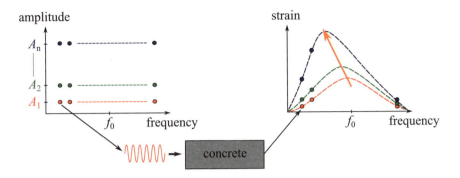

Fig. 7.55 Principle of the measure of non-linear resonance

Non-linear resonance is often used to evaluate non-classic non-linearity. The purpose is to make a given geometry test specimen resonate by changing the level of strains (Fig. 7.55).

A rapid dynamic phenomenon makes the resonance frequency decrease linearly with the amplitude of distortion. Under the 1D approximation, the slope of the line gives the non-classic non-linear parameter, which can show damage, for example.

At the LNCD lab, research is currently on-going to develop the transition to a 3D reality, giving absolute values for non-linear parameters and making a local measurement of this non-linearity possible. Time reversal is part of the methods used (see next 7.5.2).

7.5.2 Time reversal

Time reversal, which consists in a wave returning to its source, was applied to ultrasound by M. Fink (ESPCI). A time reversal experiment needs an array of transmitters/receivers connected to a multi-toned electronic device enabling the recording and re-emission of long signals.

The incident field transmitted from the source is recorded on a network of transducers, and the signals are time reversed and returned from the network of transducers in order to receive the field at the position of the initial source.

The time reversal technique is a very efficient way of focusing a wave toward its origin point by back-propagating it. There are many applications of time reversal, for example, in fields of telecommunications, medicine (lithotripsy), acoustics, and optics. In NDT, this ability is used to focus on an unknown defect, which is first subjected to ultrasound and then becomes the source.

The power of focusing can be used on any zone of a material, or even on a crack (unknown) to stimulate local non-linearity. The benefit is that the non-linear regime (high amplitude) is reached only at the very moment and position of the focus. As the reception can be carried out without contact via a laser doppler vibrometer, the point of measurement can be positioned anywhere.

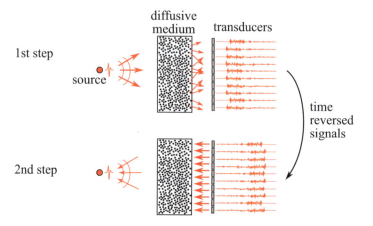

Fig. 7.56 Principle of time reversal

It is therefore possible to focus the ultrasonic energy anywhere on the specimen, the size of the focal zone in depth, and at the surface depending on the wavelength. By changing frequencies, the wavelength changes, too, as does the size of the focal zone, which enables inspection at different depths.

By concept, this focal zone can be moved, and its size can be chosen at the desired location, the one the laser points to. Applications on the search for corrosion and carbonation on concrete structures are currently being developed in the LCND laboratory.

7.6   REFERENCES

BALAYSSAC J.P., GARNIER V., VILLAIN G., SBARTAÏ Z.M., DEROBERT X., PIWAKOWSKI B., BREYSSE D., SALIN J., An Overview of 14 Years of French Collaborative Projects for the Characterization of Concrete Properties by Combining NDT Methods, NDT-CE conference, Berlin, 2015.

BANKS B., OLDFIELD G.E., RAWDING H., *La détection ultrasonique des défauts dans les métaux*, Editions Eyrolles, 1965.

BAUDOT A., MOYSAN J, PAYAN C., YLLA N., GALY J., VERNERET B., BAILLARD A., Improving Adhesion Strength Analysis by the Combination of Ultrasonic and Mechanical Tests on Single-Lap Joints, *Journal of Adhesion*, Vol. 90, Issue: 5-6, 2014: 555–568.

BODIAN P., GUY P., CHASSIGNOLE B., DUPONT O., DOUDET L., Évaluation non destructive des propriétés d'élasticité et d'atténuation ultrasonore dans des matériaux anisotropes, 10ᵉ Congrès Français d'Acoustique SFA, Lyon, April 2010.

BRUNEAU M., POTEL C., *Matériaux et acoustique, Propagation des ondes acoustiques*, Tome 1 and Tome 2, Hermes Science Publications, 2006.

BRUNEAU M., POTEL C., *Matériaux et acoustique, Caractérisation des matériaux, contrôle non destructif et applications médicales*, Tome 3, Hermes Science Publications, 2006.

CASTAINGS M., Contrôle et évaluation non destructifs de matériaux par ondes ultrasonores guidées, HDR, Université Bordeaux, 2002.

CHATILLON S., *Etude d'un système de contrôle par ultrasons des pièces de géométrie complexe à l'aide de traducteurs contacts intelligents*, thesis, Université Paris 7, 1999.

CHAIX J.F., GARNIER V., CORNELOUP G., Concrete Damage Evolution Analysis by Backscattered Ultrasonic Waves, *Revue NDT&E*, 36, 2003: 461–469.

CHAKI S., LILLAMAND I., CORNELOUP G., WALASZECK H., Combination of Longitudinal and Transverse Ultrasonic Waves for In Situ Control of the Tightening of Bolts, *Journal of Pressure Vessel Technology*, ASME Transactions, Vol. 129, 2007: 383–390.

COFREND, *Le TOFD, Time of Flight Diffraction*, Collection Les Cahiers Techniques de la Cofrend, 2013.

GARNIER V., *Evaluation Non Destructive du béton, Contribution des méthodes acoustiques linéaires et non linéaires. Apport de la fusion de données*, HDR Aix-Marseille Université, 2010.

GONDARD C., *CND par ultrasons focalisés*, Course documents on Ultrasound level 3, INSACAST, 2009.

GUY P., Contrôles Non Destructifs par ultrasons des matériaux de structure, Course documents, INSA, LYON, 2008.

INSAVALOR, *Contrôle Non Destructif par Ultrasons niveaux 2 et 3*, Course documents, 2014.

MOORE D., ROACH D., NELSON C., *Non-destructive Inspection of Adhesive Bonds in Metal-Metal Joints*, Sandia National Laboratories Report, 2009.

PAYAN C., ULRICH T.J., LE BAS P.Y., GRIFFA M., SCHUETZ P., REMILLIEUX M.C., SALEH T.A., Probing Material Nonlinearity at Various Depths by the Use of Time Reversal Mirror, Applied Physics Letters 104 (14), 144102, 2014.

PELLETIER J.L., BOCQUET M., LE TOHIC Y., *La pratique du contrôle industriel par ultrasons*, Editions Communications Actives, 1984.

ROBERT S., ALBERTINI J., SAINT-MARTIN V., BREDIF P., CARTIER F., Méthode de focalisation en tous points pour une imagerie ultrasonore à haute résolution et temps réel, COFREND Congress, Bordeaux, 2014.

ROSE J.L., *Ultrasonic Waves in Solid Media*, Cambridge University Press, 2004.

ROYER D., DIEULESAINT E., *Ondes élastiques dans les solides - Tome 1: Propagation libre et guidée*, Edition Masson, 1996.

ROYER D., DIEULESAINT E., *Ondes élastiques dans les solides - Tome 2: Génération, interaction acousto-optique, applications*, Edition Masson, 1999.

VOILLAUME H., CAMPAGNE B., Les ultrasons laser appliqués aux matériaux composites aéronautiques, COFREND Congress, 2008.

EN 12668-1 and 2: Essais non destructifs, Caractérisation et vérification de l'appareillage de contrôle par ultrasons, Partie 1 (appareils) et 2 (traducteurs), NF Standard, 2010.

EN 14127 Essais non destructifs, Mesurage de l'épaisseur par ultrasons, NF Standard, 2011.

EN ISO 16827 Essais non destructifs, Contrôle par ultrasons, Caractérisation et dimensionnement des discontinuités, NF Standard, 2014.

EN ISO 17640, Contrôle non destructif des assemblages soudés, Contrôle par ultrasons, Techniques, niveaux d'essai et évaluation, NF Standard, 2011.

EN ISO 10863 Contrôle non destructif des assemblages soudés, Contrôle par ultrasons, Utilisation de la technique de diffraction des temps de vol (méthode TOFD), NF Standard, 2011.

EN ISO 16810 Essais non destructifs, Contrôle par ultrasons, Principes généraux, NF Standard, 2014.

EN ISO 16823 Essais non destructifs, Contrôle par ultrasons, Technique par transmission, NF Standard, 2014.

EN ISO 16828, Essais non destructifs, Contrôle par ultrasons, Technique de diffraction du temps de vol utilisée comme méthode de détection et de dimensionnement des discontinuités, NF Standard, 2014.

CHAPTER 8

EDDY CURRENT TESTING

Eddy currents (EC) have an important place in the NDT field and can be applied to all highly conductive metallic materials. They can also be applied to various industrial fields such as aeronautics, automotive, nuclear, and steel industries, etc.

A conductive part is subjected to a variable, sinusoidal, or ordinary magnetic field to make induced currents appear by electromagnetic induction. These eddy currents parallel to the surface are limited to the inducing magnetic field area and mainly flow close to the surface.

A standard penetration depth is defined and is equal to the depth for which the amplitude of the surface is divided by Napier's constant e. For non-magnetic materials ($\mu_r = 1$), this depth value is around a millimeter, but for magnetic materials ($\mu_r = 100$ or more), the depth of the inspection is small (0.1 mm approx.).

Eddy currents are influenced by many parameters, which is both an advantage and a restriction for the method. Their distribution depends on the frequency, conductivity, and magnetic permeability of the material, as well as the relative part-coil geometry. The presence of a defect, emerging or close to the surface, changes these characteristics and modifies the distribution of the induced currents.

Direct measurement of the distribution is impossible, so a search is carried out indirectly. As the induced currents have the same properties as the inducing currents, they also create a magnetic field opposed to the inducing field. The value of the resulting field represents the EC distribution, that is, the presence of defects, and can be detected at the terminals of the induction coil or by another magnetic receiver (Hall Effect probe).

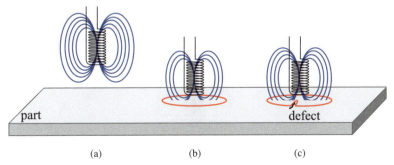

(a)   (b)   (c)

Fig. 8.1 Coil in the air (a), creation of EC (b), changed by a defect (c). The measurement is indirect via the evolution of the induction coil impedance.

8.1   PHYSICAL PRINCIPLES

Eddy currents are electrical currents induced in an electric conductor by the temporal variation of the inducing magnetic flux. Considering the geometries of media, the theoretical developments should use two-dimensional or tridimensional models. But to make it easier, unidimensional models, close enough to the reality, are used in the industrial field. They are expected to help in choosing the optimal parameters for the inspection chain.

8.1.1   Induction phenomenon in the air

A coil composed of n turns is excited by a sinusoid electric current $i(t) = i_0\cos(\omega t)$ with ω pulse. Thanks to the current, a ***magnetic field*** $H(t)$ is created in this coil and is homogenous and axial (its value is given by Ampère's circuital law). Because of the effect of this ***magnetic excitation,*** a ***magnetic induction*** $B = \mu_0 H$, is generated where μ_0 corresponds to the ***magnetic permeability of the air*** ($\mu_0 = 4\pi10^{-7}$ H/m). The current is sinusoidal, and by applying the Lenz-Faraday law of induction, we can show that an ***emf*** appears at the terminals of the coil.

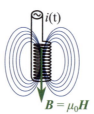

Fig. 8.2 Induction phenomenon in the air

8.1.2   Induction phenomenon in the presence of a metallic conductor

Let's imagine that the previous coil is close to an electric conductive material. A homogenous and axial magnetic field $H(t)$ is created inside the coil that is excited by a sinusoid electric current $i(t)$. In a metallic conductive material, a magnetic field $B = \mu_0 \mu_r H$ with μ_r corresponding to the ***relative magnetic permeability*** of the metallic medium is generated.

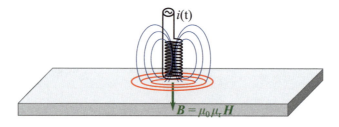

Fig. 8.3 Induction phenomenon in a conductive material

By applying the simple model of a transformer, the metallic conductor is considered to be the secondary and the induction coil, the primary. In these conditions, the magnetic flux captured by the conductor changes with time, an emf of induction appears in the material.

As the material is an electrical conductor (conductivity σ), the effect of the emf creates an induced current, the circulation of which is opposed to that of the inducing current $i(t)$ running in the coil. These eddy currents are conduction currents and create an induced magnetic field opposed to the inducing magnetic field. In these conditions, the coil captures the resulting flux and presents to its terminals a difference of potential.

The two quantities accessible to an operator are the inducing current $i(t)$ running through the coil and the difference of potential at its terminals $V(t) = V_0\cos(\omega t + \varphi)$. Module V_0 and phase φ are functions of the inducing current, the physical and geometrical properties of the material (σ, μ), the coil's geometry, and the frequency.

8.1.3 Influence of material properties and frequency on V_0 and φ

In any testing techniques, the process consists in solving an inverse problem. In EC testing, with the help of experimental measurements of module V_0 and phase φ of tension $V(t)$, we want to extract the physical and geometrical properties of the material. But first, the direct problem has to be solved. This requires connecting the behavior of the measurement parameters to the different parameters of the material.

Notion of impedance
The quantities accessible to the operator are current $i(t)$ and tension $V(t)$. Their respective representations can be illustrated in a ***Fresnel diagram***. Knowing that current i is a constant during the test, we can disregard this quantity and focus on the notion of ***electric impedance*** Z. The representation in a complex impedance plane is (Fig. 8.4):

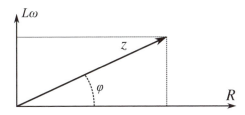

Fig. 8.4 Representation of Z in the impedance plane

Influence of the electrical conductivity of the material
By creating a hypothetical perfect coil/material coupling, the calculation of the global impedance of the assembly coil/material describes a semi-circle, a function of the parameter σ. By reasoning to the limit, when $\sigma = 0$, no EC is generated in an insulating material and the global impedance corresponds to that of the coil in the air ($L_0\omega$).

When electrical conductivity increases, EC density increases and the inductive part of the impedance (inductance) decreases. For infinite conductivity, the magnetic induced induction is equal to and opposed to the inducing induction, which gives a null total induction.

This semi-circle exists only for the perfect case previously mentioned. The curve used industrially that corresponds notably to real material thickness is orientated in a "comma" shape (Fig. 8.5), which contains the area for "impedance" vector extremities.

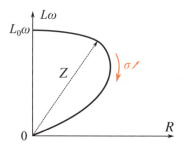

Fig. 8.6 Influence of conductivity; if it increases, inductance decreases.

Influence of material geometry

If we consider the case of a tube (or a bar) inside a circling coil, this introduces the notion of *filling coefficient* $\eta = r1/r2 = \sqrt{s1/s2}$, where $r1$ and $r2$ ($s1$ and $s2$) are respectively the radiuses (cross_sections) of the bar and the coil. Thus, the complex impedance of the coil/material assembly (Fig. 8.6) is found for different values of the filling coefficient (or magnetic coupling).

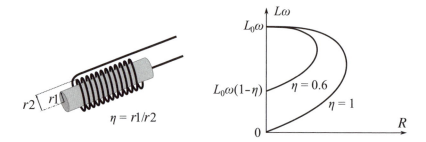

Fig. 8.5 If the filling decreases, the impedance shows a homothety equal to η.

When $\sigma = 0$, whatever the filling coefficient is, no eddy current is generated and $Z = jL_0\omega$. For all other values of σ, the resulting induction is equal to $B(S\text{-}s)$. Knowing that for a constant current impedance is proportional to the voltage, and therefore the resulting flux, its value becomes equal to $L_0\omega(1-\eta)$.

Influence of the magnetic permeability of the material

Remember that the magnetic permeability μ of the material links the magnetic field to the magnetic excitation, that is $\mathbf{B} = \mu \mathbf{H} = \mu_0 \, \mu_r \mathbf{H}$. Thus, the diagram of resulting impedance shows a homothety equal to μ_r (Fig. 8.7).

In reality, the non-linearity of the behavior of ferromagnetic materials is represented by a value μ_r, that is, a function of the magnetic excitation level. But in the particular case of eddy current testing, the use of low levels allows those non-linearity and hysteresis phenomena to be disregarded and the value μ_r to be considered as a constant.

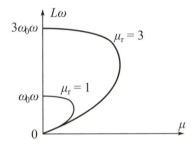

Fig. 8.7 If permeability increases, impedance shows a homothety equal to μ_r.

Influence of the frequency of the excitation signal

By definition, only the impedance of the coil depends on the frequency $Z_0 = R_0 + jL_0\omega$ with $R_0 \ll L_0\omega$. In these conditions, the scales of the impedance diagram would have to be continuously modified according to the frequency used.

As the NDT purpose is to put forward the changes in amplitude and phase of the resulting impedance (coil + material), due only to the physical and geometrical properties of the material, the impedance diagram will be standardized in order to make it independent from the electrical properties of the coil. To do so, its axes will be reduced compared to the coil in the air only. In the case of a non-ferromagnetic material, the diagram of standardized impedance (Fig. 8.8) is:

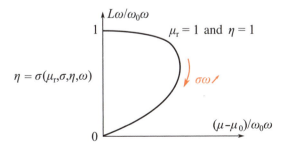

Fig. 8.8 Diagram of standardized impedance (example of a non-magnetic material)

Knowing that the emf of induction and EC are proportional to frequency and electrical conductivity, their influences are then identical on the standardized impedance diagram. Thus, the density and distribution of EC in a material are limited by the product $\sigma\omega$.

8.1.4 Penetration of eddy currents in a material

When a magnetic field is generated by a coil with a wide diameter in contact with a plane surface, a valid unidimensional model is applied. In these conditions, eddy currents are distributed in planes that are parallel to the surface. Density $J(z)$ follows a law of exponential decay, the argument of which is linked to the product $\omega\sigma\mu$, that is, to the physical properties of the material and frequency.

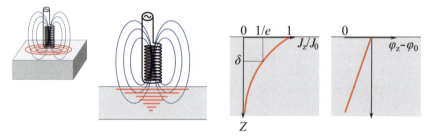

Fig. 8.9 EC penetration, conventional depth, and dephasing

The **_conventional penetration depth_** δ is then defined (Fig. 8.9) as the depth for which the surface amplitude J_0 is divided by Napier's constant e: $J = J_0/e = 0{,}368\,J_0$.

And as $J_\delta = J_0 e^{-z\sqrt{\pi f \mu \sigma}}$, we find $\delta = \dfrac{500}{\sqrt{f\sigma\mu_r}}$.

This phenomenon, called the **_skin effect_**, considerably limits the penetration depth of eddy currents in the material. There are two distinct situations regarding the magnetic properties of the conductive material:

- For non-magnetic materials, $\mu_r = 1$ and depth δ is sufficient for numerous applications, because they are in the order of a few millimeters
- For magnetic materials, μ_r is very high, and depth δ is likely to be too small (the order of a few 1/10 mm) to test for defects other than emerging cracks exclusively

A solution, which consists in magnetizing to saturation by a continuous auxiliary field, highly reduces material permeability and allows a much deeper conventional depth to be reached. For example, by saturating mild steel ($\mu_r = 1000$, i.e., $\delta = 0.1$ mm) in a field of a few thousand A/m, the value μ_r is brought to 100 and $\delta = 0.3$ mm.

But in the latter case a new problem appears – a typical example of the complex difficulties in NDT. For example, incipient cracks can be detected in a ferromagnetic tube, but it is difficult to know if these defects are located in internal or external walls. It is impossible, after saturation, to have a differentiated indication from a sufficient dephasing (Fig. 8.9) between EC outside the tube and internal EC.

The diagnosis is not accurate and appropriate remedies will not be found. As a consequence, eddy current technology is still not often used in the industrial field to identify defects in ferromagnetic tubes, whereas this technology is efficient for non-magnetic materials like austenitic stainless steels (§ 8.4.3).

8.2   TECHNOLOGY AND EQUIPMENT

An experimental device has to meet two requirements: the creation of eddy currents in the material to be tested and the detection and measurement of the variations in amplitude and phase in the resulting impedance. To obtain high sensitivity (enabling the detection of small-sized defects), the device has to be able to analyze relative variations in the order of 10^{-6} of the electrical impedance or the voltage present at the coil terminals.

8.2.1   Sensors

In all types of testing, the sensor is the most difficult element to define because the quality of the test is highly dependent on its choice.

Geometrical shape
The geometrical shape of the sensor has to be compatible with the orientation of the defect. Indeed, the circulation of EC is conditioned by the circulation of the inducing current. For sensitivity reasons, the current lines have to be normal to the main direction of the defects. Furthermore, whatever the type of sensor, the magnetic sensor/part coupling has to be kept constant, as does their relative geometrical positions.

Two types of sensors (Fig. 8.10) are then defined according to the material geometry. These are encircling sensors and internal probes dedicated to the inspection of bars and tubes, which present symmetries of revolution, and specific probes for testing a defect that has been localized.

Double function or separate function sensors
The same coiling ensures the two missions of a sensor; it creates EC in the inspected material and detects their variations in the resulting magnetic field. This is the double function sensor that measures the resulting electric impedance. This sensor is compatible with the majority of standard inspections.

However, if measurement conditions have to be optimized in order to improve, for example, the sensitivity of the test channel, these two functions are not compatible with the same coiling. Thus, the use of a separate function sensor will be favored. In emission, the coiling will have to present low electric impedance offering a maximum

Fig. 8.10 Encircling, internal, and specific probes, and a set of gauges (SOFRANEL)

level of inducing current. This will be carried out by a large diameter wire. In reception, the coiling does not generate current and only the electric voltage (proportional to the number of turns) is measured. The number of turns has to be high and made with thin wire for size reasons.

In this latter case, in order to guarantee the reproducibility of the testing conditions and EC uniformity, the two elements have to be connected together, integrated in the same box, and the emitting coil installed on both sides of the receiving element.

Absolute or differential mode
Whichever sensor is used, the mode of exploration depends on the inspection purposes.

- If the inspection results are to determine the absolute values of the physical or geometrical parameters of the material (electrical conductivity, thickness, etc.), the ***absolute mode*** is used. Voltage V at the terminals of the sensor is used to extract the absolute value of the sought-after parameter.
- If the purpose of the inspection is to reveal the presence of a large defect or to compare the physical or geometrical properties of similar materials, the variation of the measured signal is much lower compared to its nominal value. To amplify the efficient component of the signal, it is possible to use a compensated absolute mode, in which two identical sensors inspect, on one hand,

a reference sample and on the other hand, the material in question, these two materials having identical geometries and close microstructures.

- If the inspection needs a high sensitivity to show specific discontinuities, the compensated absolute mode cannot be used because the resulting voltage will be higher than the component due to specific defects. To avoid this problem, the reference coil joins the control coil in the space on the inspection sample. A *differential mode* (Fig. 8.11) is thus created and permanently analyzes the two adjacent zones of the inspected material.

The experimental device in differential mode is very sensitive in order to find small discontinuities. But it is not suitable for the detection of longitudinal defects because of the rebalancing that will occur between the two coils. In these conditions, the introduction of external or internal punctual rotating probes recreates the imbalance. Another technique consists in using one of the two coils in absolute mode, sensitive to large defects.

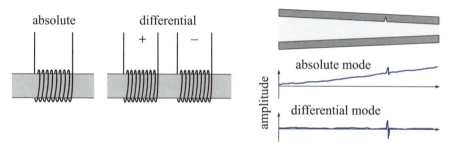

Fig. 8.11 Principle of absolute and differential modes, and real signals

8.2.2 Electric generator

An electric generator sends the inducing current through the emitting coil. Whether it is mono- or multi-frequency, the electric generator frequency is imposed either by EC penetration depth or by phase discrimination of the signals coming from discontinuities of different natures.

8.2.3 Balancing

The use of compensated absolute mode or differential mode assemblies allows partial or total independence from the physical and geometrical properties of the standard material to inspect. Thus, the measurement becomes much more accurate.

This approach stays valid only if the different coils involved in the process have equal nominal electrical impedances. In reality, although close, these impedances are not equal: balancing is needed and consists in compensating this inequality by a digital process of acquisition of the imbalance signal obtained in the absence of defect and by constantly subtracting it from itself.

8.2.4   The circuit of analysis

From the signal of imbalance coming from the inspections in compensated absolute mode or differential modes, the circuit of analysis has to extract the module ΔV and the phase φ representative of the extent and nature of the discontinuity.

8.3   METHODS OF TESTING

Today's representation of testing is carried out with the help of a digital screen representing the cathodic screen version. Horizontal and vertical deflections are linked respectively to the R and $L\omega$ voltages obtained after analysis.

In the absence of discontinuity, $R = L\omega = 0$ and the representative point of the testing is located, after balancing, at the origin (usually at the center of the screen). In the case of a discontinuity, the values of components R and $L\omega$ change and deflect the "electronic beam," which converges to another point.

This deflection is typical of the result for a specific testing method. The identification of the defect is notably possible, within a certain limit.

8.3.1   Applications to NDT: identification of the defect

The purpose of NDT is to use tools that detect and locate then identify and size a discontinuity. Whatever their nature and whether they are caused by production or damage, all defects can be represented by different types.

These defects will appear on the standardized impedance diagram, for a given material or test frequency, showing the variation in amplitude and phase of the end of the impedance vector obtained on a sound part (functioning point C).

These types of defects will be viewed by the coil (Fig. 8.12) as:
- A variation in filling coefficient η: this is the case for an external defect (chip) or a variation of geometrical dimension of the part. In this case, the evolution of the end of the impedance vector will be represented by C_1 or C_2 if η decreases or increases (on the contrary, the air gap increases or decreases).
- A variation in electrical conductivity σ: this is the case for an internal defect. The evolution will be represented by C_3 or C_4 if the conductivity decreases or increases.
- A variation in permeability μ, which is also the case for an internal defect. Here the evolution will be represented by C_5 because permeability, when it changes, can only increase.
- A coupling variation in the filling coefficient and conductivity: this is the case for cracks. The evolution will be represented by C_6 and a combination of C_1 and C_3.

Note that the speed of movement of the impedance vector's extremity characterizes the length of the defect in relation to the inspection speed. A projection of this movement on the X-axis or the Y-axis allows the implementation of X,t or Y,t "plotter type" recordings that are more relevant.

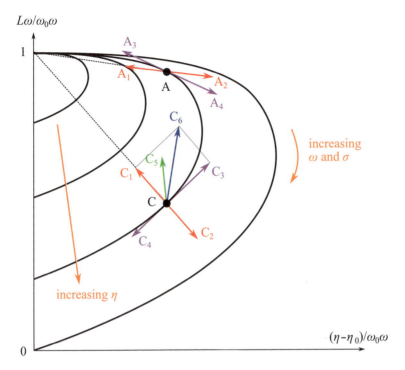

Fig. 8.12 Diagram of standardized impedance in the case of a coil surrounding a non-magnetic cylindrical conductor. Analysis of different testing cases.

The different existing theoretical diagrams (magnetic or not, cylinder, plate, etc.) will help carry out this analysis or even plan the most efficient adjustments in order to properly differentiate two types of potential defects.

Thus, the previously mentioned point C, obtained with a work frequency higher than that of point A (Fig. 8.12), is a functioning point that differentiates the defects. The external defect C_1 (or C_2) results from a 90° variation in the internal defect C_3 (or C_4). While in the case of functioning point A, the two defects A_1 and A_2 would be distinguished from A_3 and A_4 defects with variations of 20–30°, which can be declared insufficient, because it leaves too much ambiguity. Specific testing procedures recommend a minimal dephasing of 90° or even 120°.

8.3.2 Multi-frequency method

This method is used when the information resulting from a defect is likely to be hidden by the information coming from the geometry of the part. Two frequencies are injected into the coil and they are chosen to give:

- Two almost identical answers resulting from the geometry,
- Two different signals for the expected defect.

Multi-frequency equipment equalizes the amplitudes of the signals coming from the geometry, and then makes them coincide after a rotation in order to make a subtraction. As a consequence, the geometry signals will be canceled (Fig. 8.13), but the defect signals (still phase-shifted after the rotation which phased the geometry information) will remain and combine, and the defect can be detected.

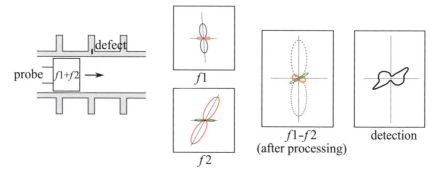

Fig. 8.13 Multi-frequency analysis

8.4   EXAMPLES OF APPLICATIONS

As we know that the response of a signal coming from an eddy current evaluation is directly linked to the physical and geometrical properties of the material, the applications can be classified in distinct fields such as:
- Dimensional inspection
- Characterization of the materials
- Soundness inspection (or defectometry)

8.4.1   Dimensional inspection

Different dimensional measures are possible, taking into account the filling coefficient. The thickness measure is key and is related to two fields of application.

Thickness measure of an (electric) conductive material
The purpose is to measure the thickness or its relative variation according to a reference (standard), by measuring the density of the eddy currents distributed in the material. The signal frequency will have to be such that the conventional penetration depth δ is higher than the thickness of the sample (Fig. 8.14).

The most current application is in the detection of corrosion in the internal walls of tubes or pipelines (external probe) or in external walls (internal probe). This thickness inspection can also be carried out by ultrasonic methods; the choice between the two techniques will be made according to the materials (conducting or not), the thicknesses (in the order of a millimeter or more), and the surface conditions (the coupling gel used in ultrasound allowing higher roughness).

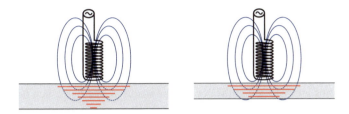

Fig. 8.14 Thickness measure of the electric conductor

Thickness measure of an insulating coating on a conductive substrate

The measurement principle (Fig. 8.15) quantifies the "lift" linked to the filling coefficient η. To increase the dynamic of the measurement, a high frequency should be chosen so that the differential between a null and infinite thickness is as large as possible on the standardized impedance diagram.

An important application field is the measurement of paint thickness, particularly in aeronautics. But this principle is also used to evaluate the presence of a metallic object and its possible movement according to the probe (position sensor).

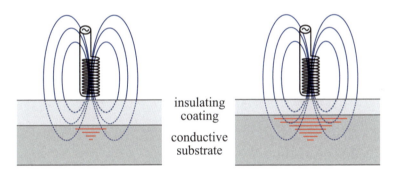

insulating coating

conductive substrate

Fig. 8.15 Measuring the thickness of an insulating coating on an electric conductor

8.4.2 Characterization of materials

While a structure is being manufactured, it is indispensable to ensure that the grades and mechanical properties of the materials of the assembly are met. Eddy currents are an appropriate means of investigation to meet these requirements. Electrical conductivity (Fig. 8.16a) and magnetic permeability are directly linked to stoichiometric concentrations and microstructures. The method used will be the absolute mode, compensated with appropriate reference samples and using double function sensors.

Sorting of different grades

It is often vital, during the elaboration phase of a structure, to make sure that the basic materials meet the requirements of the product specifications. The grades of the different alloys are compared to known references (Fig. 8.16b). These reference samples must have geometrical shapes similar to those of the materials being evaluated. As these properties are detected at the core of the material, the frequency field used will be quite low.

In the case of ferromagnetic structures such as steel, the frequency generally applied will be very low in order to take into account magnetic permeability (in the order of 100 Hz).

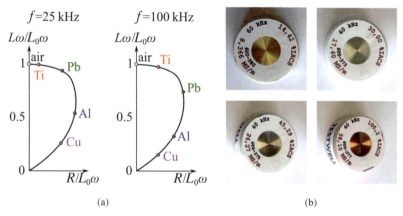

(a)   (b)

Fig. 8.16 Influence of the nature of the material on electrical conductivity (a), references for electrical conductivity (b) (TESTWELL)

Sorting of heat treatment

Heat treating materials gives them mechanical properties in accordance with the requirements of use. These heat treatments, inside or on the surface, change the crystallographic structures and microstructures of the materials, which has an effect on their physical properties. After annealing, conductivity is quite good whereas after quenching it is reduced. This is also the case after work hardening, where the dislocations that are present also reduce conductivity.

The depth of the heat treatment (quenching, for example) is also accessible by adapting the work frequencies at the levels required by the situation.

Relations with material hardness have been established where a material (steel) is as hard as its structure is thin (martensite). So, the measure of electrical conductivity by EC, via a specific calibration, gives a totally non-destructive measure of hardness compared to the hardness testing by penetration of Rockwell, Brinell, or Vickers types.

Sorting of electrical conductivities

This follows the same principle as in grade sorting but the purpose is slightly different. The density and distribution of eddy currents in the matter are directly linked to the electrical conductivity of the material. But all the eddy currents must be distributed throughout the structure.

The frequency will be chosen so that the standard penetration depth δ is much lower than the thickness of the sample. Generally, three frequencies are used (60 kHz, 120 kHz, and 280 kHz) according to the order of magnitude of the electrical conductivity of the material and its thickness. One application concerns the measurement of graphite conductivity, which checks the homogeneity of large structures such as graphite electrodes.

8.4.3   Soundness inspection

Soundness inspection, or defectometry, consists in detecting and analyzing all local modifications of the distribution of eddy currents in a conductive structure. In order to obtain appropriate sensitivity in measurement and detection, the differential mode should be used to present specific defects.

Three types of frequency ranges are used:
- LF   1 Hz to a few Hz   for ferromagnetic materials
- MF   1 kHz to 1 MHz   the most commonly used
- HF   > 1 MHz   for surface micro-cracks

Depending on the position and the nature of the defects to detect, different fields of application can be distinguished.

Detection of surface cracks

The purpose is to detect surface cracks, so the frequency field is high and a double function sensor operating in a differential mode is used. The geometry of its coils is chosen so that the presumed orientation of the cracks disrupts the circulation of the induced currents lines, as in magnetic testing (Fig. 8.17).

(a)	(b)

Fig. 8.17 Detection of well-orientated cracks, comparison between EC (a) and magnetic testing (b)

There are several applications, notably in aeronautics, with testing for possible cracks on aircraft wheel rims (Fig. 8.18), or in bolted or riveted areas (Fig. 8.19).

Fig. 8.18 Testing an airplane (or helicopter) wheel rim. The Veescan system carries out the test while the wheel is spinning, based on the principle of a record player (SOFRANEL).

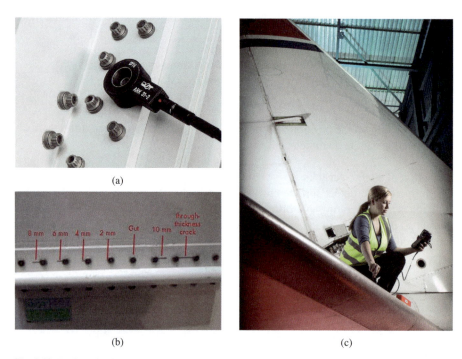

(a)

(b)   (c)

Fig. 8.19 Testing of bolted areas and reference defects on a riveted part (a) (b) (ROHMANN) and testing on-site (c) (SOFRANEL)

Testing of bored holes

Bored holes are numerous in the aeronautic sector, and eddy currents are used to detect cracks due to fatigue. These defects are generally longitudinal, so using a differential internal probe is inappropriate because it cannot show an imbalance under each coil.

Fig. 8.20 Turning probe (ETHER NDE)

Testing is carried out at high frequencies with the help of a rotating probe (Fig. 8.20) composed of two specific sensors rotating mechanically to inspect the circumference of a bored hole. A longitudinal movement is associated in order to cover the height of the hole.

As a mechanical movement such as rotation adds noise, the latest developments replace this rotation with a homogenous distribution of specific sensors on the circumference with delayed electric excitations (EC array).

Tube and bar testing

These structures presenting revolution symmetries are generally tested by differential internal or external probes (Fig. 8.21). The frequency domain depends on the

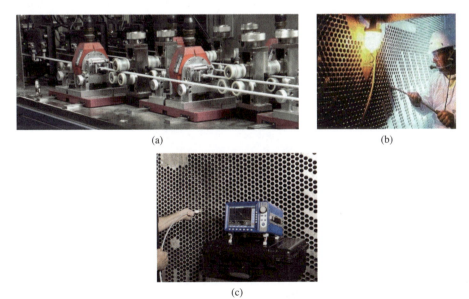

(a)

(b)

(c)

Fig. 8.21 Tube inspection from the outside during manufacturing (a) (PRUFTECHNIK) and from the inside in maintenance (b) (J.GONYEAU) and (c) (ALTA TECHNOLOGICA)

desired penetration depth in the bar or the thickness of the tube. To identify indications and obtain a relevant diagnosis, eddy currents with a 90° phase shift between the internal and external skin are created (§ 8.1.4). One of the most important applications concerns the analysis of steam generators in nuclear power plants with the help of internal probes.

8.5   CURRENT AND FUTURE POTENTIAL

Eddy current testing is possible on all conductive parts. Penetration goes deeper (a few mm) if the material is non-magnetic, and much less so (a few 1/10 mm) if the material is magnetic. In this case, the defects essentially have to be emerging or close to the surface. A defect can also be located without using any complex methods, in the tube's thickness, for example (via phase shift) and identified by considering the direction of the measured variations.

The major benefit of the EC method is the speed of testing, due notably to the fact that there is no need of a contact between the sensor and the part. Furthermore, the types of detected defects are numerous: including everything that has an influence on the distribution of EC, such as the variations of magnetic permeability, electrical conductivity, and air gap. This inspection method is used for vital parts as well as for the manufacture testing of standard parts.

Major inconveniences mostly come from the abstract nature of the electromagnetic phenomena, that is, the difficulty in interpreting results and their explanation. Note the importance of experimentation as opposed to a purely hypothetical approach, which is almost impossible.

Another important inconvenience is the non-representativeness of reference defects. An artificial crack (electro-eroded, for example) has a different volume compared to a real crack, which gives a different variation in the impedance plane. The sizing of real detected defects remains a problem even if some research has succeeded in linking the depths of certain artificial cracks (Fig. 8.22) to variations in the impedance plane.

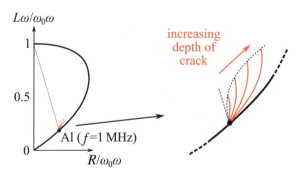

Fig. 8.22 Influence of the depth of emerging artificial cracks

There are other inconveniences, such as local variations in permeability, as in austenoferritic steels testing, and/or the "lift-off" found in automatic inspection of bars that causes a "noise" at the level of indications and weakens the signal-to-noise ratio (SNR). This problem, specific to EC is also common to several other NDT methods and will be outlined in chapter 13.

Regarding current developments in the EC method, some of these concern the automatic processing of test results (diagnosis assistance), particularly in the case of the quantity of data obtained after testing steam generator tubes. The expert systems used for the automatic interpretation of tube defects seem to be well adapted to eddy current testing, which needs both theoretical knowledge and quite a lot of practical experience.

The technique of pulsed eddy current is increasingly important. The pulse aspect enables the use of high level currents and penetrates more deeply into the material (to the order of 10 mm). As time is becoming a characteristic datum, a more selective extraction of information resulting from different depths is possible.

EC testing usually provides information such as a signal. A representation of these signals in "blankets" has been used to synthesize all the responses, but this is of no help for an analysis. Currently, the representation of the result as an image offers new solutions in the improvement of the diagnosis, by enabling the use of image processing methods. This mode of representation is also suitable for multi-element technology, which, like ultrasonic testing, can handle 32, 64, or more coils (Fig. 8.23).

The eddy current method consists in observing the intensity variations of induced currents in a conductive part. The applicability to composite materials shows that induced currents are measurable for work frequencies ranging from 1 to 30 MHz in woven reinforcements and from 10 to 500 MHz for unidirectional reinforcements.

Fig. 8.23 Cracking by corrosion under stress where the indications are represented through penetrant testing (OLYMPUS patent)

Furthermore, the difference in electric resistance between carbon fibers and polymers leads to the fact that electrically CFRP are highly anisotropic, the electric conduction of the fibers (conductor) being higher than that of the matrix (insulator). Using this anisotropy makes the determination of the fiber orientation (2D in the plane of the plies) possible for depths that do not exceed 2 mm (Fig. 8.24).

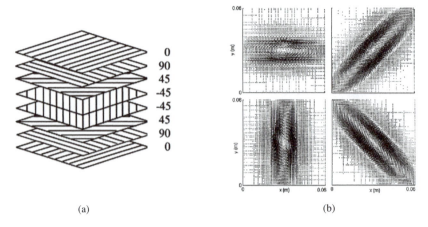

(a)   (b)

Fig. 8.24 Orientation of the plies in CFRP (a) and images of EC densities (b) (MENANA)

8.6 REFERENCES

BECKER R, BETZOLD K, BONESS K.D., COLLINS R., HOLTZ C.C., SIMKIN J., The Modelling of Electrical Current NDT Methods its Application to Weld Testing (part 1), *British Journal of NDT*, September 1986: 286–294.

BECKER R, BETZOLD K, BONESS K.D., COLLINS R., HOLTZ C.C., SIMKIN J., The Modelling of Electrical Current NDT Methods its Application to Weld Testing (part 2), *British Journal of NDT*, September 1986: 361–370.

DESCOMBESM., JAYET Y., NOËL F., Potentialités des courants de Foucault très hautes fréquences, COFREND Congress on Nondestructive Testing, Dunkirk, May 2011.

DESCOMBES M., NOËL F., ABBOUD B., ROCHER A., Appareillages et capteurs courants de Foucault hyperfréquences: principes, mise en œuvre, techniques de mesure, exemples d'application, COFREND Congress on Nondestructive Testing, Nantes, 2005.

DODD C.V., DEDDS W.E., Analytical Solution to Eddy Current Probe Coil Problem, *J. Of Applied Physic*, Vol. 6, 1968: 2829–2839.

LEBRUN B, JAYET Y, BABOUX JC, Pulsed Eddy Currents; Application to Deep Crack Detection in AU4G1, *Material Evaluation*, Nov. 95: 1296–1300.

LEBRUN B, JAYET Y, BABOUX J.C, Pulsed Eddy Current Signal Analysis; Application to the Experimental Detection and Characterisation of Deep Flaws in Highly Conductive Material, *N.D.T.& E International*, Vol. 33, no. 3, 1997: 163–170.

LORD W., PALANISAMY R., Development of Theoretical Models for Non-Destructive Testing Eddy Current Phenomena, *NDT International*, Vol. 23, 1, 1990: 11–18.

LORD R., NATH S., SHIN Y.K., YOU Z., Electromagnetic Methods of Defect Detection, *IEEE Transaction on Magnetics*, Vol. 26, 5, 1990: 2070–2075.

MENANA H., *Modélisation 3D simplifiée pour l'évaluation non destructive des matériaux composites anisotropies*, Thesis, Université de Nantes, November 10, 2009.

ROSE C., A Numerical Two Dimensional Modelling of Eddy Currents: An Application to Non-Destructive Test, *IEE Transaction on Magnetics*, Vol. 20, 5, 1984: 1983–1985.

SHAIRA M., GUY P., JAYET Y., DEVILLE L., COURBON J., EL GUERJOUMA R., Pulsed Eddy Current Evolution for Flaw Detection in Ferromagnetic Materials, *Journal of Advanced Sciences*, Vol. 17, nos.1/2, 2005: 105–111.

THOLLON F., LEBRUN B., BURAIS N., JAYET Y., Numerical and Experimental Study of Eddy Current Probes in NDT of Structures with Deep Flaws, *NDT&E International*, Vol. 28, 2, 1995: 97–102.

THOLLON F., LEBRUN B., BURAIS N., JAYET Y., Eddy Current NDT of Installed Fasteners in Aircraft Structures, Simulation and Experimental Results, 6th ECNDT, Nice France, October 24–28, 1994.

TIAN G.Y., SOPHIAN A., Defect Classification Using a New Feature for Pulsed Eddy Current Sensors, *NDT&E International*, 38, 2005: 77–82.

VUILLERMOZ P.L., *Les Courants de Foucault: principes, mesure et contrôle*, Editions AFNOR, 1994.

XIO-WEI DAI, REINHOLD L., Numerical Simulation of Pulsed Eddy Current Non-Destructive Testing Phenomena, *IEEE Transaction on Magnetics*, Vol. 26, 6, 1990: 3089–3096.

ZORNI C., *Contrôle non destructif par courants de Foucault de milieux ferromagnétiques: de l'expérience au modèle d'interaction*, Thesis, Université Paris Sud - Paris XI, 2012.

CHAPTER 9

ACOUSTIC EMISSION TESTING

Some irreversible phenomena such as plastic deformation, martensitic transformation (quenching), or damage by cracking are emission sources of transient elastic waves, resulting from local micro-displacements inside the material. Other processes such as corrosion can also create an acoustic emission.

All these phenomena correspond to damage in the material, including quenching, which is unwanted during a welding operation, for example. Acoustic emission testing (AE) consists in detecting those events with adapted sensors. It is used both as a standard NDT and NDE method for studying damage mechanisms or specific physical phenomena.

AE non-destructive testing is different from other NDT classic techniques notably because a signal can be obtained only if the defect progresses, which implies that the structure has to be loaded (mechanically, thermally, etc.). This testing is sometimes considered as "destructive," but seen logically, it is only destructive if the part is already damaged or soon to be damaged. On the contrary, the "non-destructive" aspect is accurate when the inspection is carried out on a sound part.

Fundamentally, this is a technique that monitors damage occurring during operation in real time. But it can also be adapted to the testing of non-loaded parts, for example, in manufacturing, when the monitored parts are mechanically stimulated (Fig. 9.1).

The AE method has issues when it comes to detecting a weak event in a possibly noisy environment. The problem is solved with the use of specific sensors that distinguish between the different emissions and signal processing.

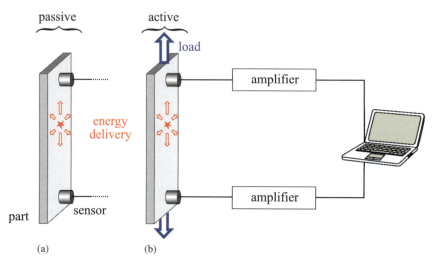

Fig. 9.1 Synoptic diagram of passive (a) and active (b) acoustic emission

9.1   PHYSICAL PRINCIPLES

9.1.1   Definition of acoustic emission

The EN 1330-9 standard of April 2000 defines *acoustic emission* as "transient elastic waves generated by the release of energy in a material or by a process." Acoustic emission is the release of elastic energy in the form of acoustic waves in a material subjected to deformation. It is the manifestation of an acoustic wave whose source is inside the material. The acoustic emission signal is an electrical signal created by a sensor as a response to these waves. Though the term acoustic emission is erroneously used, it commonly refers to the measurement technique.

These waves can be generated by numerous phenomena such as internal micro-displacements (plastic deformation, cracking), some phase transformations (martensitic), co-operative displacement of dislocations, fiber fracture, metal dissolution, release of gases (corrosion), etc.

It is a method of volume inspection, and the information obtained presents both local and global characteristics. The waves, of different nature and frequency, propagate throughout the material and are subjected to possible modifications before reaching the surface of the studied material. The surface vibration is generally collected by a piezoelectric sensor (AE frequency field ranges between 50 kHz and 1.5 MHz).

9.1.2   Acoustic emission

The acoustic emission source generates a mechanical wave and the sensor turns it into an electric signal. The way the wave propagates is largely responsible for the complexity of the signal obtained.

In an infinite solid, elastic waves propagate in two fundamental forms (cf. chapter 7 on Ultrasonic Testing): pressure (or compressional or longitudinal) waves

(P waves), and shear (or transverse) waves (S waves). They are characterized by their specific propagation velocity, the function of the medium density, and elastic constants. For example, in steel, $V_P = 5900$ m/s and $V_S = 3230$ m/s. Furthermore, in the case of an isolated source in a non-absorbing environment, the attenuation of their amplitude is inversely proportional to the distance from d to the source (decay in $1/d$).

When the solid is semi-infinite, the conditions at the limits lead to possible surface waves or Rayleigh waves (R waves). These waves are the combination of P and S, and have a lower propagation velocity ($V_R = 0.9V_S$). When the solid is a thin plate, it is also possible to obtain Lamb waves. These waves have a mode and a velocity that depend on the frequency and thickness of the plate and are referred to as dispersive waves. For these waves, coming from a specific source, the amplitude decreases to a lesser degree because in $1/\sqrt{d}$.

Thus, the attenuation of a wave, which essentially determines its detectability, is lower when this wave reaches the surface of the part surface and is turned into a Rayleigh wave – even if it is created in the volume of the part. This is a benefit for AE testing as the sensors are placed logically across the sample's surface.

9.1.3 Specific characteristics of acoustic emission testing

Important differences exist between ultrasonic testing (UT) and acoustic emission testing (AE). Generally, the number of signals obtained in AE is higher than that of UT as the emission evolves over time. The type of wave generated is more complex to predict in AE than it is in UT, even if improvements in the models used are being developed. All of this means that automated industrial systems use quite elementary descriptors.

These descriptors have to be robust enough to anticipate difficulties. As opposed to ultrasonic testing, where, generally, the complete signal is recorded, the simplified AE descriptor (such as the number of bursts in a time window) does not differentiate certain situations. This is the case, illustrated in Figure 9.2, of an event captured by two sensors, one located close to the same acoustic emission (Fig. 9.2a) and the other distant (Fig. 9.2b).

In each case, the signal is obtained in a classic way: when the amplitude exceeds a fixed threshold, the time t is noted and the digitized information between t-τ and t+T are retained, where τ and T are two predetermined values.

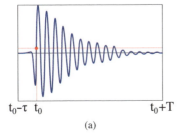 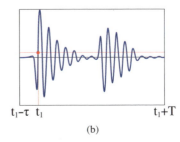

| t_0-τ t_0 | t_0+T | t_1-τ t_1 | t_1+T |

(a)   (b)

Fig. 9.2 Effects of the difference of propagation velocity on the shape of the wave: wave at emission (a) and wave after separation of P and S waves (b)

Simply counting the number of bursts does not differentiate the two situations, where one corresponds to the wave received by the closest sensor, combining P and S waves (for example, 10 alternations), from the other received by the distant sensor showing the separation of the P (fast, 5 alternations) and S (slower, 5 alternations, too).

9.1.4 Natural and stimulated acoustic emissions

Non-destructive testing by AE is different from other usual NDT techniques notably because the signal can be obtained only if the structure is stressed or loaded. But even in these conditions, only "progressing" defects can be detected. Even a large defect that does not change at all does not create any acoustic emission. The state of stress or load around the defect is essential.

If the part is naturally under stress, we refer to "natural" or "*passive*" *acoustic emission*. However, if an elastic (or thermal) stress is applied during the test, we describe it as being a "*stimulated*" or "*active*" *acoustic emission*. In this latter case, acoustic emission can be used for non-operating structures.

9.1.5 Continuous or discrete acoustic emission

Generally, the difference is made between continuous and discrete (or burst) acoustic emission.

Continuous emission (Fig. 9.3a) is characterized by the signal of an amplified noise of which the only measurable characteristic is the average amplitude. When the bursts are so frequent that they overlap each other, the acoustic emission signal shows an apparent increase in background noise. This AE is usually associated, for example, with the displacement of dislocations during the plastic deformation of metallic materials. This background noise also occurs when a leak is detected.

Discrete emission (Fig. 9.3b) is characterized by damped sinusoidal waves (*bursts*) or very short pulses. It is composed of events with an amplitude that is often more important than in the previous case. This AE is associated with damage phenomena leading to a break (initiation and propagation of cracks, corrosion under stress, etc.) and is also observed during fiber breaking.

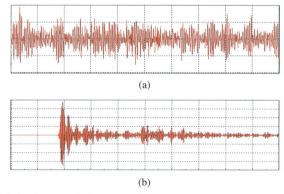

(a)

(b)

Fig. 9.3 Continuous emission (a) and discrete by burst emission (b) (MISTRAS)

9.1.6   Main acoustic emission sources

Acoustic emission sources are linked to irreversible physical phenomena and their origins are diverse:
- Plastic deformation
- Displacement of dislocations, twinning, sliding at grain boundaries, formation of Piobert-Lüders bands, breaking of inclusions or intermetallic components
- Phase transformation (martensitic, for example)
- Initiation and propagation of cracks (fatigue, corrosion under stress, etc.)
- Hydrogen embrittlement (dislocation displacement)
- Corrosion (release of gas, cracking)
- Micro and macroscopic breaking in composite materials
- Friction, mechanical impacts
- Leaks (liquid and gas), cavitation and boiling

9.1.7   Notion of source detectability (discrete AE)

The growth of a crack in a fragile material is characterized by a high propagation speed (of the crack), which gives a short signal of high amplitude. The same crack in a ductile material, however, propagates at a slower speed and is characterized by a long signal of low amplitude. Figure 9.4 illustrates this phenomenon.

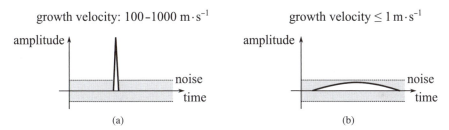

Fig. 9.4 Signal linked to the growth of a crack (a) fragile material, (b) ductile material

The detectability of a source is a function of the individual energy of each source event. The pulse of an event depends both on the volume of the created defect and the time spent on its creation. Figure 9.5 shows the consequences on the detectability threshold.

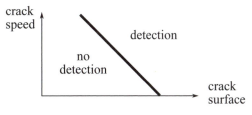

Fig. 9.5 Detectability threshold

The consequence is that some phenomena can be detected (because they can be heard over noise) and others cannot (Table 9.6). For example, the co-operative displacement (pile-up) of dislocations is the only thing that is truly detectable.

Table 9.6 Order of magnitude of the energies involved

Phenomena	Released energy (in joules)
Vibration of a dislocation	2.10^{-22} (almost undetectable)
Acceleration and deceleration of a dislocation	2.10^{-14} (almost undetectable)
Dislocation pile-up	2.10^{-8}
Grain rupture	8.10^{-9}
Test specimen rupture	2.10^{1}

So, the AE signal amplitude depends on the speed of the modification of the defect size and not on the total size of the defect only. There are some additional important factors (Table 9.7).

Fig. 9.7 Factors in the detection of acoustic activity

	Factors causing high amplitude signals	Factors causing low amplitude signals
Mechanical properties	High limit of elasticity	Low limit of elasticity
Structure	Anisotropy Presence of defects Heterogeneities Martensitic transformation Large grains	Isotropy Absence of defects Homogeneities Small grains
Rupture mode	Crack propagation Cleavage rupture	Plastic deformation Shear rupture
Changing mode	Big deformation High speed load	Small deformation Low speed load
Geometry	Large thickness	Small thickness
Environment	Low temperature	High temperature

9.1.8 Kaiser effect and the Felicity ratio

The nature of the phenomena generating AE is irreversible. The ***Kaiser effect*** refers to this property of irreversibility. It is defined as the absence of AE when a material or a structure has been carried to a certain level of load noted P1, then once it is unloaded, it no longer emits as long as the load applied during the second loading phase, remains lower than the maximal value P1 previously reached. This effect is taken into account in the inspection of pressurized vessels (§ 9.4.1) in order to "eliminate" the noise resulting from the first pressurization and then to monitor the vessel at its working pressure (lower). Figure 9.8a illustrates the Kaiser effect that occurs on all metallic materials and on some composite materials.

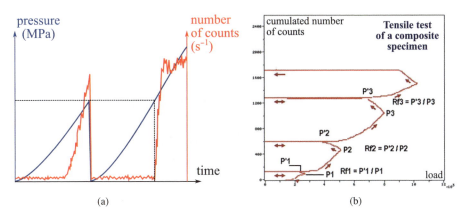

Fig. 9.8 Kaiser effect (a) and Felicity ratio obtained with a composite (b) (MISTRAS)

To detect the presence of damage, notably on a composite, the ***Felicity ratio*** can be used. This takes into account the load levels and the associated emissions. P'i refers to the level of load when the acoustic emission reappears during the $(i+1)^{th}$ loading phase Pi+1. The Felicity ratio is Rf = P'i/Pi. If P'i < Pi, it means that the structure is damaged.

9.2   TECHNOLOGY AND EQUIPMENT

The material itself generates the signal to be analyzed, which means that it is important to understand and take into account the material characteristics before considering an AE application. It can detect the evolution of a defect and even locate its position and sometimes, if the reference data are sufficient, measure its severity. However, it is difficult to characterize the geometry of a defect (depth, width, length, shape, etc.).

It is possible to test a large structure with AE in one operation only with the help of a limited number of sensors, because direct access to critical areas to inspect is not necessary.

AE is not just an NDT technique. Originally it was a monitoring technique, as the emission of stress waves is closely linked to the physical mechanisms found in the initiation and propagation of defects.

9.2.1   Equipment

The acoustic wave generated in the material has to be detected and turned into an electric signal by a sensor. This signal will be changed by amplifiers before being processed (Fig. 9.9).

Fig. 9.9 Equipment for AE testing (MISTRAS)

Sensor

The transformation of mechanical waves is carried out by the sensors. The detected signal is amplified, sampled, and stored for further processing. Knowing which type of wave is expected means that you can choose the detector that best achieves signal quality.

Piezoelectric sensors

These sensors are positioned on the surface of the material. Coupling is often ensured by a gel (silicone) in order to improve wave transmission. The sensor is attached (magnetically or in another way) and is sometimes bonded with adhesive (Fig. 9.10).

Fig. 9.10 AE sensors magnetically positioned (CEGELEC)

These are the most commonly used sensors (Table 9.11). In these sensors, a mechanical wave changes the piezoelectric crystal structure and an electric signal is generated. Lead zirconate titanate (**LZT**) is the most commonly used material. The Curie temperature of the ceramics (end of piezoelectricity) is in the order of 250°C.

These sensors are very sensitive, reliable, and stable in time, which guarantees the reproducibility of the measurements. If the sensor is correctly damped, the sensor is defined as "wide band" with an almost flat frequency response. It can then be adapted to discrete emissions, whereas a low damped sensor (resonant) will be more appropriate for continuous emissions (Table 9.11).

LZT sensors are sensitive to displacements normal to their surface. If the sensor is stimulated by a P wave moving perpendicularly to the surface, then the entire surface of the sensor will answer accordingly.

However, if the sensor is stimulated by a R wave (with a displacement parallel to the sensor's surface), the sensor will average the displacements. To solve this problem, the active surface of the sensor is reduced, but this also results in a reduction in sensitivity.

Other sensors
Optical sensors can be implemented to detect a fast displacement with low amplitude in the surface of the material to analyze. A Michelson interferometer can determine the path difference. Even if this type of sensor has numerous advantages, it is still fragile, expensive, cumbersome, and difficult to use.

A *capacitive sensor* forms a planar capacitor, one electrode of which is the surface of the part. The capacitance of such a capacitor is linked to the gap between the electrodes. This type of sensor is simple and affordable but has a limited sensitivity and is difficult to use with low conductivity materials.

PVDF (polyvinylidene fluoride) is a film that was modified during its manufacture in order to give it piezoelectric characteristics. However, it is not very sensitive. There also are piezoelectric paints prepared with ceramic powders and epoxy resin. The benefit of this paint is that it can be applied on complex shapes, but again, it is not very sensitive.

Table 9.11 Comparison of the sensitivity of a few sensors

Type of sensor	Sensitivity (in displacement)
Resonant piezoelectric sensor	10^{-13} m
Wide band piezoelectric sensor	10^{-12} m
Capacitive sensor	10^{-11} m
Interferometric laser	10^{-10} m

Calibration of the sensors, preamplifier, filter, amplifier, wire
The main characteristics of the preamplifier are the input impedance, noise, gain, bandwidth, and filter. The role of the filter is to increase the signal/noise ratio. The amplifier increases the signal close to the measurement devices so that it is easier to interpret during data analysis. The wire that connects the sensor to the preamplifier is important; it has to have a low capacitance, be shielded, and be as short as possible.

Detected signal

Coupling ensures the transmission of AE waves between structure and sensor. The characteristics of the recorded signal $V(t)$ are eventually linked to

- Source mechanisms $D(t)$
- Propagation and nature of the material (attenuation problems, echoes) $G(t)$
- Detection (resonant sensor, wide band, etc.) and acquisition system $S(t)$
- which can be summarized by: $V(t) = S(t)*G(t)*D(t)$, with convolution $*$.

9.2.2 Standards of acoustic emission sources

The first artificial source developed for the calibration of sensors was the breaking of a glass capillary. This approach was extended to the breaking of a graphite pencil lead. This solution has been chosen and standardized to simulate artificial acoustic emission sources. The break of a 2H graphite pencil lead (called Hsu Nielsen source, EN 1330 standard) generates an intense acoustic signal, similar to a natural AE source (Fig. 9.12).

This AE source is used to adjust the entire experimentation no matter the material and makes choosing the most suitable sensors possible. Furthermore, progressively breaking graphite leads further and further away from the sensor establishes a correlation between signal amplitude and distance. A correction of the amplitude measured according to the position of the source is then possible. The best listening locations for the sensors can also be determined.

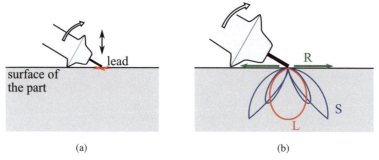

(a)   (b)

Fig. 9.12 (a) Creation of a point acoustic emission source by breaking of a pencil lead, (b) directivity of P and S waves generated by the Hsu-Nielsen source

This AE source is reproducible and quite affordable, but its weakness is its energetic nature, which is higher than the expected emissions. Recently, a ND/YAG laser has been used, which helps variation in the intensity of the simulated AE source.

9.3 DATA ANALYSIS (DISCRETE SIGNALS)

In the case of discrete signals, often used because they concern the evaluation of damage, each burst corresponds to a physical event in the material. The hypothesis is that the shape of the burst is directly linked to the characteristics of the event.

In the case of traditional acoustic emission analysis, the propagation and alterations of the signal are not taken into account. The analyzed parameters depend greatly on material properties, the geometry of the structure, the sensor, and the detection and analysis system. This analysis produces interesting (but not universal) correlations between AE parameters and sources.

For a quantitative analysis of acoustic emission, the propagation must be modeled and functions of transfer taken into account. These determine the source function independently from geometry and material properties. However, this analysis is complex, notably for a composite material, for example.

9.3.1   Signal parameters that can be exploited

The main parameters that can be exploited are represented in Figure 9.13. Most of them are defined according to a certain acquisition threshold. There are several methods for determining the threshold, the most common consists in experimentally adjusting it to a value that is slightly higher than the surrounding background noise.

Note that the threshold can be raised by 20 to 30 dB depending on where the test is carried out (laboratory, workshop, working sites, etc.)!

The classic parameters recorded in real time are:
- Peak amplitude in decibels
- Duration between the first and the last threshold overshoots
- Number of threshold crossings, also called number of counts
- Rise-time in microseconds
- Average frequency of a burst
- Signal energy
- Frequency content of the signal

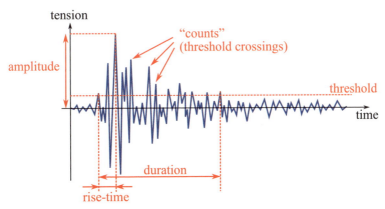

Fig. 9.13 Main parameters measured in real time on an AE signal

These parameters, except for amplitude, depend greatly on the acquisition threshold value, which will have to be carefully chosen. It can be also interesting to establish ratios in order to create new variables that can give more interesting signal representations. It is also possible to digitize the entire signal (streaming) and carry out time-frequency analysis, but the problem is that a very large amount of data can quickly result from this analysis.

9.3.2   Location of the defects

The purpose of location is to determine the point or the zone in which the AE occurred. The method can be linear, 2-dimensional, or tridimensional. Figure 9.14 shows different sensor arrangements for locating the source of samples of different geometries. A linear location on a tensile test specimen needs two sensors only. More complex geometries need a higher number of sensors.

Fig. 9.14 Different sample geometries and positioning of sensors enabling the location of AE sources

The position of the source is calculated according to the different signal time of arrival to the sensors and the wave propagation speed in the material. The location is performed thanks to ***triangulation*** (Fig. 9.15) using the arrival time on several sensors. An algorithm adapted to the geometry of the location mesh calculates the position of the source. To inspect large structures, many meshes are associated, which forms a real sensor network.

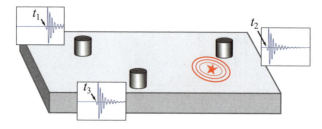

Fig. 9.15 Triangulation and location

The evaluation of wave propagation velocity can be done by using the pencil-lead breaks procedure on the surface of the part (Fig. 9.16). Another possibility consists in using the signal attenuation law and combining it to a location mesh, provided that the studied material is known. For composites, which are heterogeneous, and anisotropic materials, the diffusion and the dispersion of acoustic waves are so important that the specific location of a source becomes difficult.

This velocity anisotropy makes the location of the source difficult. It is easier to first do it on a zone by only focusing on arrival times of the waves to the sensors. Once the zone of interest is located, a finer triangulation can possibly be carried out.

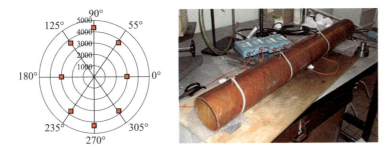

Fig. 9.16 Propagation velocity of the waves (m/s), measured with the pencil-lead breaks procedure on a glass/epoxy tube with crossed plies (–/+ 55°) (INSA)

9.3.3   Chronological or time analysis

The evolution of one or many AE parameters is studied according to time or any other test parameter (stress, displacement, temperature, etc.). This is the most commonly used analysis method and is qualitatively excellent and relatively easy to implement. For example, the chronology of damage on concrete can be defined during a tensile test. A typical result is shown in Figure 9.17.

The curve presents three characteristic zones of acoustic emissions associated with crack creation and propagation. In zone A, there are zero to few acoustic emissions. This zone corresponds to the linear deformation of the material. In zone B, AE signals of low amplitude start to appear.

It is generally considered that in this area the formation of microcracks or the stable propagation of cracks occur. A sudden increase of AE that proceeds together with signals of much higher energy is a sign of unstable propagation of cracks (coalescence) and the ruin of the material (C zone).

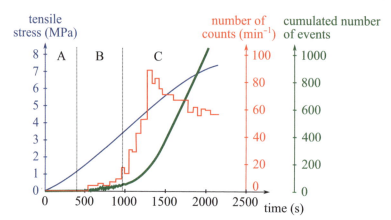

Fig. 9.17 Typical acoustic activity during a tensile test on concrete

9.3.4   Statistical analysis

This consists in drawing cumulated distribution curves and histograms of the distribution of AE events according to any one of the parameters. The most commonly
used analysis is the distribution of peak amplitude. It gives quite a good idea of
the detectability of the acoustic emission and the influence of the threshold on the
recorded activity.

9.3.5   Analysis of correlation

In *correlation* analysis, a parameter is plotted according to another. It is then possible
to plot the number of counts versus the signal duration, energy versus amplitude, etc.,
which makes separation into different signals groups possible. This analysis is used to
eliminate unwanted signals (mechanical, hydraulic noises, etc.).

9.3.6   Multi-parameter analysis

The previous methods analyze pairs of different acoustic parameters, but although it
is possible to display several windows during the test, the analysis of results is long
and sometimes difficult. More powerful *multi-parameters analysis* techniques are
developing and more rapidly detect significant parameters.

    Diverse mathematical methods enable a statistical data analysis which deter-
mines similarities or differences, by analyzing not one single characteristic parameter but n parameters, also called descriptors. The extraction of class recognition
criteria from large quantities of data is the fundamental principle of this analysis. For example, Fisher's method and *principal component analysis* (**PCA**) are
counted among these techniques, together with k-means clustering and artificial
neural networks.

9.3.7   Decomposition of the signal into wavelets

The *Fourier transform* has long played a major role in the analysis and processing of signals. The spectrum obtained gives the frequency composition of the original signal, but not when these frequencies have been emitted. The wavelets are a recent method of signal analysis. A *wavelet* replaces the Fourier transform sinusoid by a translation-expansion set of the same function. The result can be expressed as a time-frequency representation that is different and complementary, enabling a temporal vision of the frequency distribution.

9.4   EXAMPLES OF INDUSTRIAL APPLICATIONS

9.4.1   Requalification of pressurized devices

Acoustic emission has two major benefits for the *requalification* of pressurized devices (Fig. 9.18):
- It guarantees the safety of a *pneumatic test* by detecting during the test any defect evolution likely to cause an unexpected breaking of the device
- It helps estimate the device lifetime

Fig. 9.18 Requalification of a storage sphere (a) (GROUPE INSTITUT DE SOUDURE) and gas cylinders (b) (MISTRAS)

Taking into account the Kaiser effect, three different cases can occur:
- The absence of AE: the structure is in the same state as in the commissioning phase; there is no creation or evolution of defects;
- The detection of AE: as soon as the stress exceeds the working pressure it means that there are working defects in evolution; and
- The detection of AE at a pressure lower than working pressure: this means that the origin of AE is not the evolution of a defect (friction, etc.) because the Kaiser effect is respected, and the device has already been subjected to a higher stress in operation.

9.4.2 Monitoring

There are several fields of application corresponding to investigation, inspection, or
structure monitoring methods:

- Monitoring in operation: AE is the only technique able to detect the creation
  of defects and monitor their propagation. It is used to inspect pressure devices
  when they are tested and during operation (Fig. 9.19).

Fig. 9.19 Installation of AE sensors on oxygen vessels (and zoomed image) (GROUPE INSTITUT DE
SOUDURE)

- Inspection of buried LPG tanks: AE integrates all stresses inherent to these
  containers – inaccessibility of welding, the need to carry out the test onsite.
  The test is a propane pressurization test. Leak detection is based on the analysis
  of the variation of the efficient value of the signal.
- Airframe monitoring.
- Manufacturing monitoring: AE can detect anomalies in real time according to
  the required manufacturing conditions – spot welding test, material production, quality of press-forming, good manufacturing of composite and wooden
  structures, etc. (Fig. 9.20).
- Detection of leaks: a leak actually causes an increased background noise; it is
  a continuous type emission.

Fig. 9.20 Stimulated acoustic emission on a wooden airplane wing (MISTRAS)

9.4.3 Evaluation of the level of damage on concrete

The use of the Felicity ratio and AE during the loading and unloading phases of a tensile test set up at increasing maximal load on refractory concrete gives an idea of how damaged a structure is.

Some authors characterize damage on reinforced concrete beams with parameters such as "load ratio" and "calm ratio." "Load ratio" is the load value for which the acoustic activity starts again compared to the maximal load of the previous cycle. This ratio corresponds to the Felicity ratio. "Calm ratio" is the cumulated acoustic activity during unloading compared to the total acoustic activity cumulated at the maximum of the previous cycle. These two parameters draw a map of the level of damage in the studied structure.

9.4.4 Other characterizations

AE is an additional investigation method used in laboratories that study materials. It gives information about the internal structure of the materials and their deformation mechanisms because the waves are generated after their internal modifications. This technique is, for example, used to monitor mechanical tests (Fig. 9.21) (fracture, fatigue), study damage mechanisms, and develop treatments (thermal, surface, etc.).

Although acoustic emission seems to be the only technique able to detect all sources of damage in composite materials, the interpretation of information contained in AE data requires a careful analysis of damage to the material and the collected signals.

Fig. 9.21 AE tools for test specimens subjected to tension (INSA)

9.5   CURRENT AND FUTURE POTENTIAL

This is an NDT method that can evaluate a structure in real time. Monitoring is continuous with a high sensitivity. It is also the only non-destructive method that enables the inspection of two scales at the same time, a global metric scale in relation to the entire structure (a tank for example), and a local millimeter scale when location is possible.

In the first case, the purpose is to detect in-service defects in industrial applications and in the second case to characterize materials and structures in the laboratory.

The difficulty in modeling the signal obtained comes from the lack of prior knowledge about the type of waves generated by the source, the waveform which will propagate in the structure, and the acoustic characteristics supplied by the sensor and the recording system.

It is difficult to differentiate the noise from the AE signal, and users have to learn from their own experience. Moreover, quantitative correlation is limited to current experience because it is very hard to transpose to other situations.

Current works essentially concern the modeling of the sources and the consequences on the signal obtained (direct problem), but also the recognition of phenomena from collected information (inverse problem). The complete digitization of the signal is sometimes necessary, and research is focusing on fast algorithms for data analysis.

The next paragraphs outline a few examples of current further improvements.

9.5.1   Directivity of the sources

Various works have calculated the spatial distribution of the waves emitted around a mode I (microcracking) or a mode II source (shearing). Figure 9.22 represents, for a crack, the amplitude of P and S waves emitted in all directions of the plane. These theoretical results have been subjected to an experimental check on a CT (compact tension) specimen enabling the positioning of several sensors in different directions of space that simultaneously received the emitted waves from the same source event.

It is then possible to differentiate the directional properties of a source by measuring the P and S amplitude ratio.

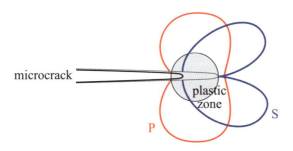

Fig. 9.22 Directivity of the P and S waves associated with a growing microcrack

9.5.2   Prediction on material lifetime

The purpose of this approach is to evaluate the residual lifetime of a material, that is, its *remaining potential*. The idea is to predict the lifetime of a working part by analyzing its behavior at the beginning of loading.

The damaging events occurring at the beginning of service life have a major influence on the lifetime of a part. These are called precursory events. Knowing about these events, notably by AE, which records the bursts corresponding to precursory events, enables the evaluation of the remaining potential.

Disorder is an inevitable concept if the nature of real materials has to be taken into account. Initial damage is often diffuse and not correlated before it becomes organized. Disorder is a key element in the rate of events precursory to macroscopic fracture. A fracture can be predicted only by looking at the co-operative aspect.

Focusing on the evolution of acoustic activity in chronological or temporal analysis can be one approach. A significant increase in acoustic activity is looked for, for example, during a fatigue test where an inflection point (called Ti) appears on the cumulated curve of acoustic events in a function of time (Fig. 9.23).

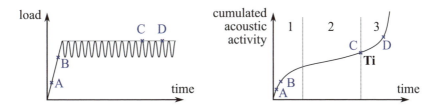

Fig. 9.23 Evolution of acoustic activity during a fatigue test

This point is always associated with the beginning of the concentration of damage in the test specimen, close to the breaking zone. It is systematically present and always appears "a certain time" before the breaking of the material. It could then be used to predict the remaining lifetime.

9.6 REFERENCES

BEATTIE A.G., Acoustic Emission, Principles and Instrumentation, *Journal of Acoustic Emission*, Vol. 2, no. 1/2, 1983: 95–128.

EITZEN D.G., WADLEY H.N., Acoustic Emission: Establishing the Fundamentals, *Journal of Research of the National Bureau of Standards*, Vol. 89, no. 1, 1984: 75–100.

GODIN N., HUGUET S., GAERTNER R., Influence of Hydrolytic Ageing on the Acoustic Emission Signatures of Damage Mechanisms Occurring during Tensile Tests on Unidirectional Glass/Polyester Composites: Application of a Kohonen's Map, *Composite Structures*, 72, 2006: 79–85.

GROSSE C., OHTSU M., *Acoustic Emission Testing*, Editions Springer, 2008.

MAILLET E., GODIN N., R'MILI M., REYNAUD P., FANTOZZI G., LAMON J., Real-time Evaluation of Energy Attenuation: A Novel Approach to Acoustic Emission Analysis for Damage Monitoring of Ceramic Matrix Composites, *Journal of European Ceramic Society*, Vol. 34, Issue 7, 2014: 1673–1679.

MOMON S., MOEVUS M., GODIN N., R'MILI M., REYNAUD P., FANTOZZI G., FAYOLLE G., Acoustic Emission and Lifetime Prediction during Static Fatigue Tests on Ceramic-Matrix-Composite at High Temperature under Air, *Composites Part A: Applied Science and Manufacturing*, Vol. 41, Issue 7, July 2010: 913–918.

MORIZET N., GODIN N., TANG J., MAILLET E., FREGONESE M., NORMAND B., Classification of Acoustic Emission Signals Using Wavelets and Random Forests: Application to Localized Corrosion, *Mechanical Systems and Signal Processing*, Vols 70–71, March 2016: 1026–1037.

KISHI T., OHTSU M., YUYAMA S., *Acoustic Emission beyond the Millennium*, Editions Elsevier, 2000.

RIETHMULLER M., L'émission acoustique: applications aux équipements industriels, COFREND Congress, 2008.

ROGET J., *Essais non destructifs. L'émission acoustique: Mise en œuvre et applications*. Editions AFNOR-CETIM, 1988.

ROUCHIER S., FORAY G., GODIN N., WOLOSZYN M., ROUX J.J., Damage Monitoring in Fibre Reinforced Mortar by Combined Digital Image Correlation and Acoustic Emission, *Construction and Building Materials*, Vol. 38, January 2013: 371–380.

SOTIRIOS J., VAHAVIOLOS J., *Acoustic Emission, Standards and Technology Update*, ASTM Editions, 1999.

CHAPTER 10

INFRARED THERMOGRAPHIC TESTING

Infrared thermography is a method by which an object can be inspected remotely and without contact via the imaging of its infrared radiation.

Any matter, living or not, radiates an energy depending on its temperature. Industrial temperatures correspond to infrared radiation. A camera, sensitive to this type of radiation, detects it. The optical flow is then converted into an electric signal that is in turn converted into a visible thermal image in a color scale.

Any defect adding a difference in the distribution of the radiation is visualized. This method is used when the materials to inspect are naturally hot (degradation of the refractory coating of a furnace, monitoring of installation insulation, etc.) or cold (CO_2 containers), but also in the context of electrical installation maintenance (high voltage power lines, transformers), measurement of glass temperature, and manufacture conditions of windshields, bottles, fibers, etc. In this case, we talk about passive infrared thermography (Fig. 10.1).

It is also possible to thermally excite the material when the searched defect is likely to disrupt the flow and then be detected. We talk about active thermography, in reflection or transmission, often used in the inspection of composite materials.

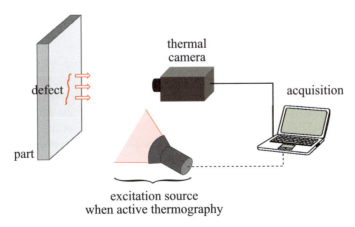

Fig. 10.1 Synoptic of passive and active thermographic testing

10.1 PHYSICAL PRINCIPLES

10.1.1 Emissivity of the materials

The interaction of an electromagnetic radiation of wavelength λ with a material causes three mechanisms (Fig. 10.2):
- Absorption $\alpha(\lambda)$ and re-emission $\varepsilon(\lambda)$ by radiation;
- Transmission $\tau(\lambda)$, which depends on the radiative properties of the material: an opaque body has a null transmission coefficient (for a certain range of wavelength) – there are transparent and semitransparent bodies; and
- Reflection $\rho(\lambda)$ in the direction of the incident flux but in the opposite direction.

The absorbed energy warms the object until the thermodynamic balances with the external environment, then the object re-emits as much energy as it has absorbed. *Emissivity* is the physical quantity that allows us to know the quantity of re-emitted flow by radiation after absorption. In the industrial temperature field (usually ranging from $-30°C$ to $2000°C$), the re-emitted radiation $\varepsilon(\lambda)$ is essentially of an ***infrared*** type.

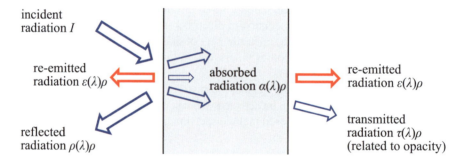

Fig. 10.2 Radiation-material interactions

Luminance is defined as the intensity of emitted radiation per unit of surface perpendicularly to the initial direction of the emission (chapter 3). Infrared thermography thus measures the instant distribution of luminance of the surface of an object in the sensitivity spectrum of the camera. This "qualitative" distribution is either sufficient to establish a NDT diagnosis or the distribution of the real temperatures need to be "quantified" and determined in respect to laws adapted to each material.

10.1.2 Emissivity of a black body

A ***black body***, an ideal material, is used as a reference because its radiated spectrum depends only on its temperature. It is a perfect absorber of all the incident radiation and a perfect emitter (there is no reflection $\rho(\lambda)$, no transmission $\tau(\lambda)$). Thus, a black body emits the maximum of thermal energy for a given temperature and wavelength whereas real bodies will emit less.

The behavior of real bodies is different from that of a black body. They absorb a part of the radiation, reflect another part, and transmit the rest. It is then more difficult to link the luminance to the temperature. They are linked via emissivity $\varepsilon(\lambda)$, a dimensionless quantity equal to the ratio of the energy radiated by a material (its luminance) and the energy radiated by a black body ($\varepsilon = 1$) at the same temperature (so, $\varepsilon < 1$).

On the other hand, the more a material is black (which spectrally means that visible waves are absorbed) and dull (low reflection), the closer to 1 emissivity is, whereas the more reflective a body is, the lower the emissivity.

Table 10.3 The emissivity of some materials

Material	Emissivity (for $\lambda = 8$ to $14 \ \mu m$)
Theoretic black body	1
Human skin	0.98
Concrete	0.95
Water	0.93
Graphite, carbon	0.7–0.9
Oxidized steel	0.7–0.9
Stainless steel, Inconel	0.1
Polished aluminum	0.05
Polished gold	0.02

10.1.3 Infrared radiation

Planck's law gives the luminance of a black body according to its temperature, and *Wien's law* gives the wavelength λ_{max} at which the luminance of a black body is maximal for a given temperature (in °K):

$$\lambda_{max} = 2898/T \quad \text{in } \mu m$$

- for ambient T 20°C λ_{max} 10 μm
- for 500–600°C λ_{max} 3 μm
- for 2000°C λ_{max} 1 μm

These wavelengths correspond to infrared radiation (Fig. 10.4), and detection devices must have this sensitivity.

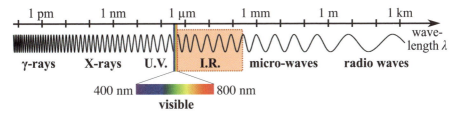

Fig. 10.4 Ranking of electromagnetic radiation

If the wavelength of infrared radiation is high enough, it will have a low energy W, because $W = hc/\lambda$ with h = Planck constant, and c = electromagnetic wave velocity. Thus, the measurement seems to be delicate.

10.1.4 Real materials

Emissivity depends on the material, temperature, and wavelength considered, but also on its surface condition and the direction of emission (Fig. 10.5). Thus emissivity is maximal for a normal incidence (0°) and decreases strongly for an incidence higher than 55°. The incident angle therefore has to be maintained close to 0°.

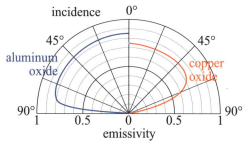

Fig. 10.5 Influence of incidence angle on the directional total emissivity, using the example of aluminum and copper oxides

10.1.5 Transmission through the atmosphere

Transmission through the atmosphere greatly disrupts the measurement. But as it is effective in the 1.9–2.5 μm, 3–5 μm, and 8–14 μm bands, due to a lower absorption, infrared thermography cameras operate in these three windows of transmission (Fig. 10.6).

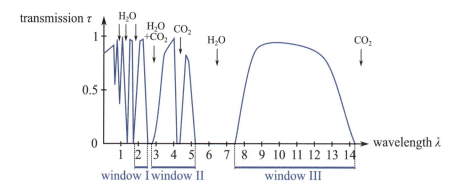

Fig. 10.6 Transmission coefficients τ of infrared in the atmosphere are important in three ranges of wavelength λ

10.1.6 Radiometric equation

The camera gives a thermal image that is not quantified in temperature. To obtain a real thermogram, the measurement of the radiation coming from an object has to be accompanied by an analysis in order to determine the behavior of the real object observed compared to the "ideal" matter and to determine the part of the measured radiation truly emitted by the object compared to the reflected part. However, this observation is not sufficient to quantify or analyze the temperatures (in the profession, we say that the camera measures the radiation but it is definitely the expert, the thermographer, who evaluates the temperatures).

A ***radiometric equation*** defines the energy measured by the camera. In a given direction and subject to simplifying hypotheses (environment is considered as a black body, object assumed opaque, i.e., no transmission, proximity of the object, etc.), it is possible to write

$$I_{\text{measured}} = \varepsilon I_{\text{black body}}\,(T_{\text{object}}) + (1-\varepsilon)I_{\text{black body}}\,(T_{\text{environment}}).$$

In practice, when it is not possible to put a thermometer in contact for calibration, abaci are used to calculate the temperature of an object in relation to an ideal black body. For example, a real body heated to x degrees, with an emissivity of 0.65, shows the camera a luminance of 105 cd·m^{-2}. The emissivity correction gives 105/0.65 = 160 cd·m^{-2} as the luminance of a black body heated at this temperature, that is, $x = 100°C$ according to Figure 10.7.

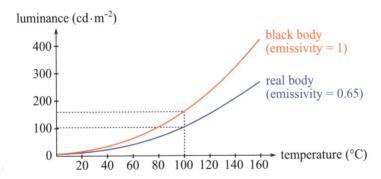

Fig. 10.7 Black body – real body correspondence and influence of emissivity

In addition to emissivity (or the ability of the material to emit a radiation), the type and reflection factors, and the temperature of surrounding objects, the thermographer sometimes needs to consider the transmission factor of the observed materials.

These materials have to be opaque to the radiation to which the camera is sensitive, that is, they have to stop the propagation at their surface. Several traps can be found with transparent and semi-transparent materials, such as glasses, plastics, and paints. Some "noise" coming from the environment located behind the non-opaque object can also occur.

10.2 TECHNOLOGY AND EQUIPMENT

10.2.1 History

Between 1900 and 1920, numerous patents were registered for devices that detected persons, artillery, planes, ships, and even icebergs. The interwar period saw the development of image converters that literally allowed an observer to see in the dark.

However, sensitivity was limited to close infrared waves, and targets had to be illuminated by infrared search beams ("active" thermography), which ran the risk of being seen by the enemy.

After the war, military research programs worked to develop passive systems, but military secrecy prevented any other developments until the mid-1950s. In medical practice, breast cancer detection is the oldest application, but radiography and other methods eventually turned out to be more efficient. The first infrared camera for industrial use was released on the market in 1968 by Swedish company AGA/Bofors to find defects in electrical lines.

10.2.2 Infrared thermography cameras

Nowadays, there are three types of camera on the market:
- SW (short wave), which measures the emitted radiation in the 1–2 μm spectral band;
- MW (medium wave) in the 2–5 μm spectral band; and
- LW (long wave) in the 8–12 μm spectral band. These are favored for long-distance measurements because transmission through the atmosphere is better.

These cameras are currently composed of detector matrices, either thermal (they heat under radiation and some of their physical properties change according to the heating) or photonic (they turn the received energy into electricity). In the latter case, performance is better but they need to be cooled after use.

Thermal or infrared cameras (Fig. 10.8) are defined by:
- Their spatial resolution (the smallest visible object)
- Their thermal resolution (the smallest temperature difference between the bars of a pattern, at a given frequency perceptible by the eye on the visualizing screen)

These two resolutions are not independent, and the cameras are characterized by the curve representing the evolution of the thermal resolution in relation to the space resolution. This curve is called the MRTD (minimum resolvable temperature difference).

    (a)   (b)   (c)

Fig. 10.8 Portable infrared thermography cameras (a) (Fluke) (b) (Flir) (c) (Testo)

The thermal sensitivity of a camera is the NETD (noise equivalent temperature difference), which is the smallest detectable temperature difference. A low NETD gives a better detection of thermal contrasts. The required minimum for an operator in thermography is of 0.1°C, but cameras can be found on the market with a NETD lower than 0.045°C.

10.2.3 Image quality

The non-visible infrared image is turned into a visible image by assigning the radiation any level of grey or a color code which is often red for hot and blue for cold.

    The image quality obtained is a very subjective notion that integrates numerous influence parameters: resolution, sharpness, etc. The spatial resolutions of observation and measurement are, for example, not necessarily identical: a sufficient number of pixels is needed to measure and fewer to create an image.

    For example, a hot wire seen on a screen can have an insufficient diameter to be properly measured in temperature. A particular degradation of this quality, the signal-to-noise ratio, comes from the transmission through the atmosphere, even if the camera spectral windows are adapted.

10.2.4 Specific technological needs: military, medical, industrial

Table 10.9 A few examples of applications in infrared thermography (Photos LAVOISIER*)

Application field		Military	Commercial		
		Environment (+ security)	Industry	Medical	
General applications		Navigation reconnaissance Night flights Target acquisition, shot	Measure of pollution, natural resources, reduction of energy (+ police, firemen)	Maintenance manufacturing process inspection, NDT	Pathologies of vascularization Fever detection
Specific needs	Sight stabilization	Necessary	Useless in general		
	Image processing	Detection and automatic target recognition	Optional		
	Resolution	Target at a long distance or detail in visual field	The distance is adaptable		
	Duration of processing	Real time	Real time not always necessary		
	Sensitivity	At the perception limit (low NETD)	Contrast often sufficient (NETD not dominant)		

10.3 MÉTHODS OF TESTING

10.3.1 Passive infrared thermography

Any defect that adds a difference in the distribution of emitted radiation by a mechanical system can potentially be detected by infrared thermography.

For example, this method is used for the monitoring of installation insulation or in the context of preventive maintenance of mechanical installations, etc.

The studied temperature field is usually in the order of −30°C to +2000°C, with adaptable scales (every 100°C or every 0.1°C).

10.3.2 Active infrared thermography

It is possible to thermally excite the material when a defect does not "naturally" add a difference in the distribution of emitted infrared radiation. Active thermography is different from the passive method as it uses a controlled external thermal source in order to create an artificial temperature contrast between sound areas and areas containing a defect. This method requires a prior check to ensure that the searched defect disrupts the flow and that there is detection potential.

Stimulations used can be direct (lamp, flash, etc.) or indirect (ultrasound, induction). A part is locally heated and the evolution of the cooling is analyzed. Cooling can indicate degradation, notably in the case of composite materials: lack of adhesive on sandwich structures, delamination on monolithic composites, etc.

In active thermography, as the stimulation source characteristics are known and controlled, it is even possible to model the inspection and obtain quantitative results (position, quantity, etc.) on the heterogeneities detected in the material.

The homogeneity of the heating is then an important criterion of the method's success in order to properly link the variation of the thermal flow to a particular defect.

10.3.3 Heat sources in active infrared thermography

Different heating sources are used, particularly according to the physical characteristics of the material to inspect: optical methods (halogen lamps, flash lamps, laser), electromagnetic methods for conductive materials (transmitted currents, induced currents), or acoustic methods (ultrasound).

Sometimes natural heating (sun) is sufficient to notice, for example, leaks in hydraulic ducts (aerial view before and after sunset). Heating with a warming blanket is used to detect possible water locks in certain plane wings. The use of hot pressurized air is also possible.

We talk about *pulsed thermography* (Fig. 10.10) when the propagation of the heat front and the temperature differences that a defect will create are observed. *Modulated thermography* consists in studying the dephasing and the amplitude of the temperature's response in the part in relation to the sinusoidal thermal stress in a steady state. *Pulsed phase thermography* combines the two and the benefits are a fast execution for the first, whereas the second is freed from the non-uniformity of heating.

The use of ultrasound (*vibrothermography*) produces heat for defects presenting contact surfaces (crack, delamination, etc.). The part is not heated and the problem of the heating homogeneity is therefore less important. The principle is similar when the Joule's effect produced at the level of defects by the disruption of an eddy current field is used.

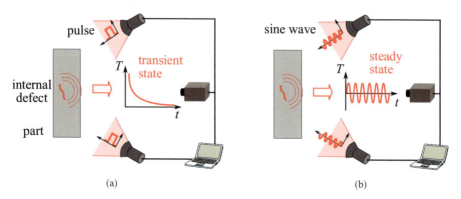

Fig. 10.10 Pulsed thermography (a) and modulated thermography (b)

10.3.4 Thermography: NDT and/or NDE method

Infrared thermography is a particular non-destructive method with multiple applications in military, medical, and industrial fields. It enables NDT to search for cracks on steel walls or delamination in composite materials, etc., both in an active or passive way.

It also enables the user to carry out very specific NDE, such as the characterization of materials and structures (in a broad sense, for example, a flume, the Earth, etc.), the only condition being whether the searched parameters introduce or disrupt temperature fields.

10.4 EXAMPLES OF APPLICATIONS

10.4.1 Passive thermography

These kinds of applications historically justified the use of infrared thermography in the industrial field.

Monitoring of internal refractory coating degradation
Kiln refractory coatings are subject to degrading during use. In the case of continuous working, a solution consists in regularly stopping the installation and changing the coating. There is no optimization. With infrared thermography, it is possible to adapt maintenance to real fatigue conditions. Figure 10.11 shows two examples of possible degradation of the internal coating.

Monitoring of building insulation
An experimental standard was written in 1983 for building inspection (Fig. 10.12). It was approved in 1999 (EN 13187). In September 2012, there was still no regulation imposing thermography as an observation technique and method of temperature measurement in buildings.

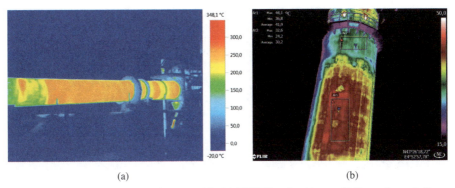

(a)   (b)

Fig. 10.11 Infrared thermography of a cement kiln (a) (TESTO) and a chimney (b) (image from a helicopter, Cie de L'Etang)

Fig. 10.12 Building insulation inspection (IUSTI)

Monitoring of insulation and lagging of industrial systems

Thermography can be used to check the lagging systems of pipes and other containers. It can also be used for the possible discoveries of hot (Fig. 10.13 and 10.14) and cold thermal bridges (Fig. 10.15). When iced areas are found, it is important to pay attention to possible heating and the appearance of melting ice causing corrosion problems.

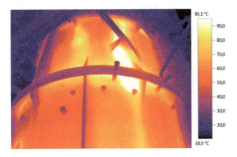

Fig. 10.13 Lagging monitoring on a vertical kiln (TESTO)

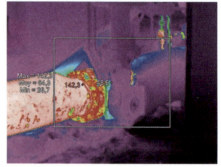

Fig. 10.14 Lagging defects with visible and thermographic images (SGS)

Fig. 10.15 A cold thermal bridge on a cold-insulated exchanger with visible and thermographic images (KEM ONE)

Level measurements

The temperature difference can evaluate the level of hot or cold fluids (Fig. 10.16).

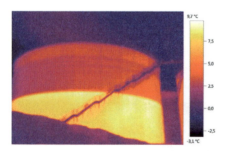

Fig. 10.16 Measurement of the level of liquid in storage containers (TESTO)

Inspection of local damage

Local damage (cracking or piercing, collapsing or hollowing, etc.) in an assembly (kiln, pipeline, chimney, etc.) containing a fluid at a different temperature from the outside will cause a variation (positive or negative) in the local temperature (cf. Fig. 10.17) and can be seen with infrared thermography.

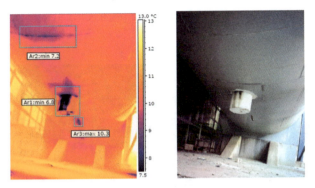

Fig. 10.17 Inspection of container and tapping with suspicion of a local defect (KEM ONE)

Monitoring and maintenance of electrical installations

Infrared thermography helps visualize the thermal signatures associated with an electric malfunction causing a temperature rise (Fig. 10.18). Even if the measured temperature is the surface temperature, the inspection is very quick, notably when it concerns complex electric panels.

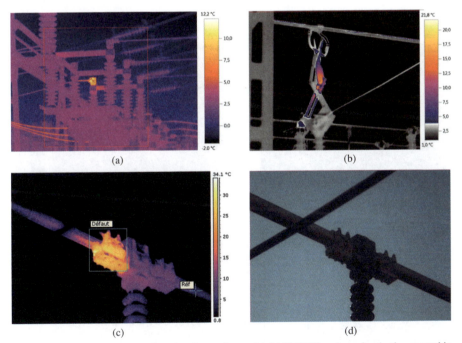

Fig. 10.18 Defects on elements of an electric transformer (a) (b) (TESTO) and on a bar (c, thermographic image) (d, visible image) (RTE)

In the context of high-voltage line monitoring, a camera installed on a helicopter (Fig. 10.19) or a drone can give a quick view of local heating showing malfunctions.

Human intervention on a line is necessary only if the detected temperature exceeds a certain reference value.

Fig. 10.19 Helicopter and drone (RTE) used for high-voltage line inspection (TESTO)

Checking manufacturing temperatures

Thermography is particularly efficient for the monitoring of glass manufacturing (Fig. 10.20). It can also be used for other fabrications (Fig. 10.21), fast or not, for which a temperature variation would be synonymous with a defect.

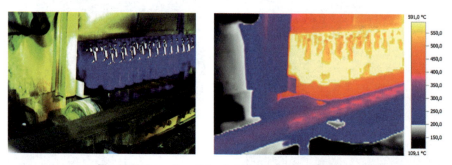

Fig. 10.20 Inspection of glass bottle manufacturing (TESTO)

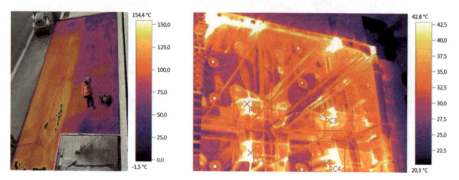

Fig. 10.21 Inspection of bituminous mix deposit on a road (viewed from above) and of an operating plastic injection mold (zoomed-in view) (TESTO)

Monitoring maintenance of mechanical installations

A defective mechanism can cause abnormal heating (Fig. 10.22).

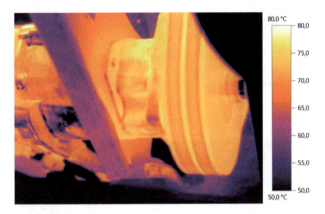

Fig. 10.22 Losses caused by heating in power transmissions (TESTO)

Temporal monitoring of a thermal phenomenon

Passive infrared thermography studies the evolution of thermal phenomenon over time. It is complementary to many works on the characterization of materials (Fig. 10.23).

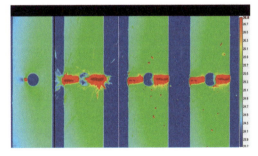

Fig. 10.23 Evolution of the cooling of a cup of coffee related to its material (after 1 min. and 9 mins) and progressive heating (at 0 ms, 20 ms, 40 ms, and 60 ms) caused by the breaking of a composite material (IUSTI)

10.4.2 Active or stimulated thermography: application to soundness inspection of composite materials or sandwich structures

The surface temperature of a material is measured during excitation (low, as it has to be non-destructive!) and during the consecutive cooling (Fig. 10.24).

This method is used to detect defects (air bubbles, composition inhomogeneity, delamination, cracks, etc.) in materials such as composites, as their low thermal conductivity gives time to visualize the phenomena. The detection of the defect is shown by an inhomogeneity of the surface temperature (Fig. 10.25).

Fig. 10.24 Installations of stimulated infrared thermography (line method) on a sandwich panel and a Rafale part (DASSAULT AVIATION)

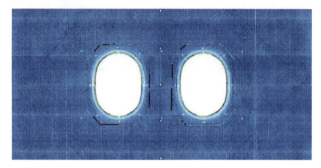

Fig. 10.25 Micro-delaminations (in black) detected on a bent sandwich composite panel (DASSAULT AVIATION) (cf. their macrographs Fig. 2.11a)

The thermophysical properties of a material can also be characterized. The evolution of the surface temperature of a material over time can be compared to a 1D analytical model of heat transfer, which enables, by using an inverse method, the estimation of conductivity and diffusivity of the material. The method is currently applied to the characterization of thermophysical properties of polymers and composites.

The high anisotropy of the thermal properties of a honeycomb composite board is taken into account in the reference thermal image. The presence of a defect will disrupt the balance (Fig. 10.26), in particular the lack of adhesive on the two interfaces.

(a)   (b)

Fig. 10.26 Separations shown (in red and yellow) in a sandwich panel, foam and fiberglass skin (a), and at the skin-honeycomb interface (b) (DASSAULT AVIATION)

10.5 CURRENT AND FUTURE POTENTIAL

Infrared thermography is a very efficient NDT and NDE method, notably when the temperatures of the installations to inspect are high. In the active version, the field of application concerns increasingly different situations where materials and structures are at ambient temperature.

In all cases, for NDT the suspected defect has to be likely to change the distribution in the temperature fields. Information about the structures is then accessible, and some people use this technique for parts as different as integrated circuits and concrete bridge decks.

Thermography can be carried out on parts with complex geometries in a wide range of temperatures with a 0.1°C resolution or even lower. Just like (local) ultrasonic testing, access to one side of the part is enough, but this time the inspection is global. Implementation can be simple with material that is easily transported.

The disadvantages of the method come from variations in the surface emissivity, from possible heterogeneity of heating and therefore from a difficult calibration. Measurement is very difficult (almost impossible) if the atmosphere (which ensures the transmission of radiation) is disrupted by dust or thermal flows (the sun).

A rapid decay in the method's relevance can be noticed according to the depth of defects: it is essentially a sub-surface technique.

When the problems are solved, the images of structures or defects obtained are as relevant as with classic methods such as ultrasonic testing but with the speed of a less localized inspection.

Current improvements for passive thermography essentially concern the equipment: better transportability, better resolution, more efficient image processing. But for active thermography, research on the potential of pulsed or modulated or combined part heating, or selective heating of the defects are in a consolidation phase. The study of more efficient digital processing of thermographs is also a particularly active research axis.

10.6 REFERENCES

BOUTEILLE P., LEGROS G., *La thermographie infrarouge active sur pièces de fixation mécaniques*, Editions CETIM, 2016.

CANIOU J., *L'observation et le mesurage par thermographie*, Editions AFNOR, 1991.

COFREND, *Thermographie: certification du personnel CND selon EN 473, EN 4179 et ISO 9712*, 2004.

COFREND, *Thermographie Niveaux 1-2-3*, Annales Officielles de la Certification Cofrend, 2012.

CLINIQUE ST JOSEPH LIEGE, *La thermographie infrarouge en maintenance prédictive*, case study, Editions Ministère de la Région Wallone, 1991.

GAUSSORGUES G., *La thermographie infrarouge, principes, technologies, applications*, Editions Tec et Doc, 1999.

KAMINSKI A., *La thermographie infrarouge comme technique de mesure et de contrôle non destructif*, course document, INSA Lyon, 2012.

KOUADIO T., *Thermographie infrarouge de champs ultrasonores en vue de l'évaluation et du contrôle non destructifs de matériaux composites*, Doctorate thesis, Université de Bordeaux, 2013.

LARGET M., *Contribution à l'évaluation de la dégradation du béton: thermographie infrarouge et couplage de techniques*, Doctorate thesis, Université de Bordeaux, 2011.

MALDAGUE X., *Non Destructive Evaluation of Materials by Infrared Thermography*, Editions Springer Verlag, 2012.

NF A09-420: *Essais non destructifs, Thermographie infrarouge, Caractérisation de l'appareillage*, 1993.

PAJANI D., *Mesure par thermographie Infrarouge*, ADD Editeur, 2012.

PAJANI D., *La thermographie du bâtiment: Principes et applications du diagnostic thermographique*, Editions Eyrolles, 2012.

PAPINI F., GALLET P., *Thermographie Infrarouge, Image et mesure*, Editions Masson, 1994.

RIGOLLET F., REICHLE R., GASPAR J., GARDAREIN J.L., LE NILIOT C., HUXFORD R., Prediction of Spatial Resolutions of Future IR Cameras at ITER, 11th International Conference on Quantitative InfraRed Thermography, QIRT 2012.

SAM ANG KEO, *Nouvelles Méthodes de Contrôle Non Destructif (CND) en Génie Civil: Développement de la Méthode de Thermographie Infrarouge par Excitation Micro-onde et Laser CO_2*, Editions PAF, 2014.

CHAPTER 11

LEAK TESTING

Historically, the industries that used leak testing the most were those manufacturing items with a high intrinsic hazard potential or high technological value (space, nuclear, chemistry, etc.).

This testing gained its good reputation in 1942 in the United States with the Manhattan Project and 20 years later in Pierrelatte, France, with the building of a factory for the isotopic enrichment of uranium by gaseous diffusion.

This method was then extended to non-hazardous products in the case of a leak but where non-conformity could lead to a commercial risk (for example, a small trace of oil on an automobile engine).

Here the malfunction (leak) is searched for directly. It is created by a through-wall defect, no matter the origin of this defect. Leak causes are diverse: welding defects, cracks, matter porosity, loose gaskets, scratches on seal bearings, etc.

The various methods are based on a search for a leakage flow, which most of the time corresponds to the quantity of gas going through a wall in a given time and for a given pressure difference. Tightness is then linked to the object's function: for example, different leakage flows are admitted for a compressed air cylinder or a pacemaker.

Different methods are used according to the size of the estimated leak, the shape of the part, the environment, etc. The methods include vacuum box, helium test (Fig. 11.1), pressure variation, acoustic emission, etc.

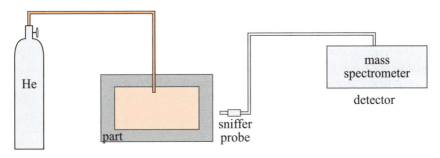

Fig. 11.1 An example of leak test using the helium sniffing principle

11.1 PHYSICAL PRINCIPLES

11.1.1 Leakage flux

A *tightness defect* refers to the absence of matter, which makes a leak possible. By definition, a *leak* is the passage of a fluid from one side of a wall to another and this can happen in different ways:

- By fluid diffusion between matter grains. We talk about *permeation*, such as the passage of helium through rubber joints or air noble gases going through the glass in lighting tubes. These leaks are the most common and are so small that they are generally tolerated.
- By fluid circulation in "through" ducts of diverse origin: scratches on seal bearing, undertightening, etc.

11.1.2 Flow

To exist, a flow q needs 3 elements:
- a fluid
- a tightness defect with dimensional characteristics C
- a pressure difference ΔP

$$q = C \, \Delta P \qquad \text{in } \mathrm{Pa \cdot m^3 \cdot s^{-1}}.$$

Leak flow is influenced by:
- the nature of the fluid (viscosity, molar mass);
- the temperature, in operation or during the inspection;
- the pressure to which the tightness defect is subjected; and
- the environment of the defect influencing the ability of the passage of the gas stream.

Leak flow measurement is the direct consequence of the control of the parameters above. And in addition, you also need to be aware of the risk of a temporary blocking of the leak.

11.1.3 Unit

The $\mathrm{Pa \cdot m^3 \cdot s^{-1}}$ is the IS (International System) unit and must be used instead of former units ($\mathrm{mbar \cdot l \cdot s^{-1}}$, $\mathrm{torr \cdot l \cdot s^{-1}}$, lusec, etc.).

If a leak is considered as a straight tubular duct of 1 cm long subjected to a pressure difference of 1 bar, the flows correspond to:
- A tube diameter of 10 μmeter for a flow of 10^{-5} $\mathrm{Pa \cdot m^3 \cdot s^{-1}}$
- A tube diameter of 0.15 μmeter for a flow of 10^{-11} $\mathrm{Pa \cdot m^3 \cdot s^{-1}}$

11.1.4 Thresholds

A leak flow of 10^{-4} $\mathrm{Pa \cdot m^3 \cdot s^{-1}}$ corresponds to a loss of 5000 Pa in 10 hours in a volume of 10 liters. This is an important leak for a compressed air cylinder, but a 10^{-10} $\mathrm{Pa \cdot m^3 \cdot s^{-1}}$ leak corresponding to a loss of 1 $\mathrm{cm^3}$ every 30 years is also harmful if the object is a pacemaker.

The acceptable threshold is then linked to the function of the object. The leak thresholds generally tolerated in different industries are listed in Table 11.2.

Table 11.2 Leak thresholds in different industries. Illustration of the tightness need in the nuclear sector, for the protection of the operator and for tubular beams.

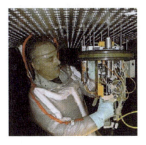

Automotive	approx. 10^{-5} Pa $\cdot$ m$^3 \cdot$ s^{-1}
Aeronautics	approx. 10^{-7} Pa $\cdot$ m$^3 \cdot$ s^{-1}
Nuclear	approx. 10^{-8} Pa $\cdot$ m$^3 \cdot$ s^{-1}
Ultra-high vacuum	approx. 10^{-11} Pa $\cdot$ m$^3 \cdot$ s^{-1}

11.2 TESTING TECHNIQUES

Leak testing is an NDT method that is generally applied to completely finished parts. They have to be clean, dry, and grease-free. The mechanical or thermal stresses have to be released before the inspection.

The EN 1779/A1 standard defines the selection criteria of a method through 17 leak testing techniques.

To implement these tests, 5 main techniques are used:
- Bubble testing
- Pressure variation testing
- Ammonia testing
- Halogen gas testing
- Helium testing by mass spectrometry

11.2.1 Bubble test

Two techniques are used according to the type of pressurization.

Part pressurization
After the blanking of unnecessary openings, the part is pressurized with air or any other gas. It is important to confirm the correct mechanical behavior of the part so a hydraulic test has to be carried out beforehand.

The part then is immersed in a liquid (usually water) for a global inspection or sprayed with a surfactant solution (soap) on the relevant zone for a local inspection. The leak is characterized by the appearance of regular *soap bubbles* or foam at the location of the through-wall defect (Fig. 11.3).

Although this inspection is not quantitative, it is possible to quantify a leak by collecting bubbles in a measuring cylinder, per time unit. The sensitivity of this technique is in the order of $1 \cdot 10^{-4}$ Pa $\cdot$ m$^3 \cdot$ s^{-1} and is a function of the pressure.

(a)   (b)

Fig. 11.3 Soap bubble test with leaking weld (a) and defects in the expanding of tubes (b) (ACTEMIUM)

Use of a vacuum box

The surface of the zone to be inspected is sprayed with a surfactant solution then a transparent **vacuum box** is applied to the surface and vacuumed to 500 mbar to locate important leaks and then to 200 mbar for thinner leaks. It is necessary to be careful about the saturation vapor pressure of the surfactant product because it is likely to generate artifacts!

This type of inspection is applicable to structures under construction. It is often used on assemblies where only one face is accessible, such as the bottom of storage containers for oil products or the metallic liner of a reactor building, etc. It is not exclusively used for closed structures, and it can be carried out on plane surfaces (Fig. 11.3).

The sensitivity of this technique is also in the order of $1 \cdot 10^{-4}$ Pa$\cdot$m$^3 \cdot$s^{-1}. The dimensional characteristics of the vacuum box have to match the geometry of inspected zones.

11.2.2 Pressure variation test

After the blanking of unnecessary openings, the part to be inspected is pressurized with a pressure different from the atmospheric pressure, either via air pressurization or any other gas (be careful about the relevance of the hydraulic test) or via a vacuum. In the presence of a leak, the internal pressure will tend to the atmospheric pressure (Fig. 11.4).

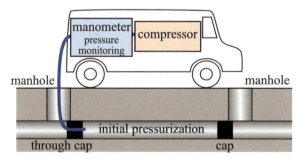

Fig. 11.4 Leak testing of urban pipes

Despite easy implementation, the final result might be doubtful because of many factors, particularly:
- Degassing from the part when tested with vacuum
- Presence of dead or spurious volume
- Reference pressure
- Temperature variations

The value of the leak is calculated via $q = -\Delta P \cdot V/\Delta t$ and you must have accurate information about part volume, measuring assembly volume, and temperature variations.

The sensitivity of the global inspection is in the order of $1 \cdot 10^{-5} \, Pa \cdot m^3 \cdot s^{-1}$ and is a function of the volume, measure duration, and sensitivity of the pressure sensor. Many fabrications of measurement systems by automated **pressure variation** are integrated into assembly lines, particularly in the pharmaceutical, automotive, and household appliance industries.

This method is particularly appropriate to all industry sectors that need to accurately guarantee the tightness of their parts:
- Automotive: injection circuits, calipers and brake systems, radiators, pumps, cooling systems, headlights, air conditioning systems, hoses
- Appliances: washing machines, refrigerator compressors, boilers
- Medical: catheters, syringes, bags, bottles, etc.

11.2.3 NH₃ ammonia test

The part to be inspected is first vacuumed (≤ 20 mbar) or inerted with nitrogen, before **ammonia** pressurization. The zone to be inspected, usually weld beads, is passivated with the help of an acid solution and dried. A reactive paint composed of bromophenol blue is applied to the weld bead.

Fig. 11.5 Leak testing with ammonia (IFAT)

With an initial pH ranging between 2.7 and 3.2, the paint is yellow (Fig. 11.5). On a through-wall defect, the ammonia gas flow generates a pH increase and the paint becomes locally blue. After inspection, the paint is brushed, the ammonia is collected and dissolved in the water. After neutralization, this solution can be disposed of.

A precise visual examination of the paint condition then validates the tightness of a part via the localization of a through-wall defect. The sensitivity of the technique is in the order of $1 \cdot 10^{-7} \, \text{Pa} \cdot \text{m}^3 \cdot \text{s}^{-1}$.

It is possible to "quantify" a leak via an empirical method using the Cortelly diagram, which takes into account the evolution of the diameter of the stain over a time unit in relation to the pressure in the part and its thickness.

It is important to remember that ammonia is a dangerous gas; it is explosive with concentrations in air between 15 and 28%. The olfactory threshold is at 5 ppm. The presence of copper or copper alloy must absolutely be avoided.

Although its use imposes restrictions, the ammonia test is still used for leak testing of weld beads on a certain type of LNG tankers, ships transporting liquefied natural gas.

11.2.4 Halogen gas test

Halogen gases used to be widely employed as tracer gases in leak testing in the 1980s in electrical industries and also for welding inspection in certain types of LNG tankers, notably because helium (a "competing" tracer gas) is particularly expensive. After the Tokyo Summit in 1992 and the Kyoto Protocol in 1997, the emission of halogen gases in the atmosphere (destroying the ozone layer) was expressly forbidden.

Nowadays, this inspection remains in the use of process halogen gases as tracer gas. From *sulfur hexafluoride* (SF_6) to *dichlorodifluoromethane* (CCL_2F_2) and from the electrical industry to that of refrigeration, this testing is implemented in different ways.

Pressurization with sniffing and accumulation

The part is pressurized via its tracer gas. It is placed either in an auxiliary container or the zones to be examined are put into airtight plastic bags. The *tracer gas*, traversing the through-wall defects, accumulates in the container (or bag) and causes an increase in concentration. The sample and the analysis are carried out by a *sniffer* probe connected to the leak detector.

According to the accumulation, duration, and volume of the container (or bag) at the atmospheric pressure, it is possible to calculate the global value of the leak, without defining the origin(s).

The sensitivity of this technique is in the order of $1 \cdot 10^{-7} \, \text{Pa} \cdot \text{m}^3 \cdot \text{s}^{-1}$. The calibration is carried out with the help of a calibrated leak connected to the part, at the same pressure and concentration conditions and flowing into a known volume.

Pressurization with direct sniffing

The part is pressurized via its tracer gas. The zones to inspect are scanned by the sniffer probe (speed ≤ 10 mm/sec), connected to the leak detector. When there is a leak and according to its amplitude, the signal is analyzed and compared to the signal issued by the calibrated leak connected to the part.

The sensitivity is in the order of $1 \cdot 10^{-7}$ Pa·m^3·s^{-1}. The implementation of the test helps locating through-wall defects, in addition to the "accumulation" technique.

11.2.5 Helium testing by mass spectrometry

Vacuum leak testing with *helium* by *mass spectrometry* has remained the most sensitive and reliable test for over 50 years. A calibration is carried out from a standard leak. The latter is composed of a helium container (Fig. 11.6) at the atmospheric pressure with, at its extremity, a porous membrane that generates a perfectly identified reference gaseous flow caused by vacuum effect. The leak amplitude ranges from $1 \cdot 10^{-10}$ to $1 \cdot 10^{-7}$ Pa·m^3·s^{-1}.

There are also (capillary) *calibrated leaks* connected to the part that gives a reference signal according to pressure and concentration. The leak amplitude ranges from $1 \cdot 10^{-7}$ to $1 \cdot 10^{-1}$ Pa·m^3·s^{-1}.

Fig. 11.6 A standard leak (PFEIFFER VACUUM)

Caution: it is very important not to mistake the internal calibration of the leak detector and the calibration of operative conditions. The vacuum calibration of operative conditions is carried out through a standard leak, which connects to the part at the furthest point from the mass spectrometer, creating a gaseous flow generating a helium signal that is stabilized via a time response.

The calibration of operative conditions, when it is pressurized, is carried out through a calibrated leak connected to the part, which generates a gaseous flow that creates a helium signal identified via a sniffer probe connected to the mass spectrometer.

One of the benefits of the helium test is also linked to the fact that this gas is not used in the degassing of tanks (Fig. 11.7). No helium remains except obviously that constant in the air (which can be used for measurement calibration). No random "background noise" is to be feared.

Fig. 11.7 Direct helium sniffing technique used for the inspection of welding in LNG tankers (ACTE-MIUM)

There are different techniques using helium tracer gas.

Global vacuum testing
The part is vacuumed after blanking of unnecessary openings. When the internal pressure is lower than 1 Pa, it is linked to the mass spectrometer. After identification of the signals related to the calibration of operative conditions (be careful about degassing and dead volumes), the part is placed in an auxiliary container or in an airtight bag (Fig. 11.8).

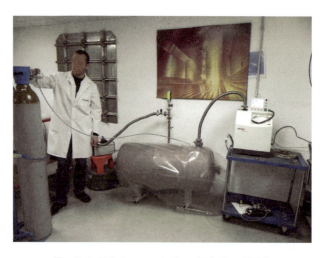

Fig. 11.8 Global vacuum testing with helium (IFAT)

Helium is then injected into the container or bag and concentration is measured. When a leak occurs, helium penetrates into the part and is analyzed and quantified by the mass spectrometer. The sensitivity of this technique is in the order of $1 \cdot 10^{-10}\, Pa \cdot m^3 \cdot s^{-1}$. Today, it is often used in the space industry and in fundamental research.

Beyond global testing, it is possible to locate the leak(s), after removal of the plastic bag or by removing the part from its auxiliary container, through a partial vacuum inspection: suspected zones (weld beads, bolted assemblies, etc.) are put in plastic bags and helium is injected. A "helium spray" test can also be carried out, where a small flow of helium is projected onto suspected zones while respecting a displacement speed of a few mm/s.

Global testing with sniffing and accumulation

Just like halogen gas testing, the part is pressurized with helium after blanking of unnecessary openings and checking of its mechanical resistance compatibility with the test, before being inserted in an auxiliary container or an airtight plastic bag. The helium traversing through-wall defects accumulates and concentration increases. According to the time of accumulation and the enclosure (or bag) volume at atmospheric pressure, it is then possible to calculate the global value of the leak without defining the origin(s). The sensitivity is in the order of $1 \cdot 10^{-7}\, Pa \cdot m^3 \cdot s^{-1}$.

Calibration is carried out either with the help of a calibrated leak connected to the part in the same pressure and concentration conditions and flowing in a known volume, or with the natural concentration of helium in the air, that is, 5.2 ppm $(5.2 \cdot 10^{-6})$.

The sample and the analysis are carried out by a sniffer probe inserted in the bag or container and connected to the mass spectrometer. Beyond the global inspection, it is possible to locate the leak(s) after removal of the plastic bag or removing the part from its auxiliary container by direct sniffing and scanning of the suspected zones using the probe (speed $\leq$ 10 mm/s).

This technique is often used for the leak testing of welds aboard operating LNG tankers and in thermal exchangers in the electronuclear field.

Vacuum box testing

The vacuum box is applied on one of the faces of the zone to be tested and then vacuumed before being connected to the mass spectrometer. The opposite face is subjected to helium. The helium seeping through the defect is analyzed and quantified by the mass spectrometer. This type of inspection is applicable on structures under construction, it is notably used on assemblies where only one face is accessible.

The sensitivity of this technique is in the order of $1 \cdot 10^{-9}\, Pa \cdot m^3 \cdot s^{-1}$ (according to the tightness of the vacuum box). The dimensional characteristics of the vacuum box have to match the geometry of the inspected zones.

Pressurization/vacuum testing

This technique is applied to small volume parts, sealed before testing. It consists in installing the part to be tested in a high-pressure helium pressurization chamber after vacuum. The time of immersion depends on the thickness of the walls, the size, and the internal volume of the leak. In order to avoid interfering memory effects, the part is aerated to let the external surfaces degas: it is then put in a vacuum chamber connected to a mass spectrometer. The helium having penetrated the part during the pressurization phase is analyzed and quantified during the external vacuum operation.

The sensitivity of the technique is in the order of $1 \cdot 10^{-9}$ Pa$\cdot$m$^3 \cdot$s^{-1}. Note that care must be taken with important defects that are likely to generate a helium signal that does not correspond to reality. This technique is used in the semiconductor and watchmaking industries (airtight/waterproof cases).

11.2.6 Acoustic emission

As already explained in chapter 9, an acoustic emission can be detected by different systems. In the present case, we look at an acoustic source that could correspond to a fluid flow through an opening, valve, or crack that causes turbulences, cavitation, vaporization, etc.

The benefit of tightness measurement using this method is that it can be carried out without interrupting production. Wave guides can be used for hot objects and for non-accessible structures as well (Fig. 11.9).

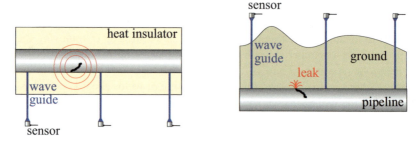

Fig. 11.9 Leak detection under insulator or on a buried pipeline using wave guides

11.3 LIMITS OF TEST EFFICIENCY

The part has to be finished, with no machining, welding, or other phases to come. In order to be detected, the tightness defects have to be open, clean, dry, and not greasy (including gaskets).

Finally, a leak test can only be properly carried out on parts that have no dead volumes within their construction (Fig. 11.9), because leaks to be detected can be so thin and small that the transfer between the outside and the inside in the case of a dead volume can last for many years.

The solution passes through discontinued welds or by machining holes connecting the dead volume with the inside of the part.

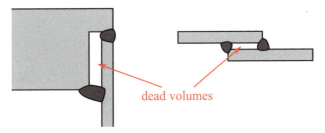

Fig. 11.10 Leak testing cannot be applied to parts with dead volume

11.4 CURRENT AND FUTURE POTENTIAL

Since the Magdeburg experiment carried out by Von Guericke using a vacuum bell without any hatch or opening established the fundamentals of tightness technology, many developments have occurred.

Leak testing via the multiple techniques that can be implemented are everywhere: in the protection of food products, the containment of hazardous materials, the "cleanliness" and purity of supports and deposits of thin coatings on the chips of electronic cards, the comfort of vehicle cabins, etc., including for products with high technological value (space industry, nuclear industry, chemistry, etc.).

The disadvantages of leak testing mostly come from the fact that the shape of real leaks is different from that of model leaks used for calculations and based on perfect capillaries. There is a gap in prediction, but practical experience seems to show that the resulting errors are within a factor of 2.

It is important to remember that leaks add up. A leak corresponding to a 0.8 mm diameter hole in a container may be harmful but can be easily found, whereas 10 leaks of 0.08 mm diameter will not necessarily be individually detected. Once again it is the responsibility of the designer to anticipate this possibility in order to define the optimal testing method.

Ensuring tightness requires knowing the rules adapted to the products to be inspected and the goal to be reached. In optimal cases, the sensitivity of helium type methods is so great that some people claim that this technique cannot be improved. Of course, helium mass spectrometers have evolved and manufacturers are focusing on simplifying their use via automation and computer science. "All gas" mass spectrometers are even used nowadays.

11.5 REFERENCES

BLANC B., HENRY R.P, LECLER CJ., *Guide de l'étanchéité, Volume 1 et 2 (les fuites)*, Edition Société Française du vide, 1992.

CASADO F., et al., Contrôle de l'Etanchéité des cuves méthaniers en construction, COFREND congress, Nice, 1990.

CASADO F., et al., Contrôle de l'étanchéité et classification des enceintes de confinement, COFREND congress, Nantes, 1997.

CASADO F., NDT on Liquid Natural Gas Carriers in Service Membranes Tanks Testing, 15th WCNDT, Rome, 2000.

CASADO F., et al., Qualification du détecteur de fuites azote hydrogéné, Congrès COFREND Dunkirk, 2011.

HENRY R.P., *Cours de science et technique du vide*, Editeurs Société Française des Ingénieurs et Techniciens du vide, 1971.

SEEMANN B., *Le contrôle d'étanchéité*, Editions Eyrolles, 2008.

TALLON J., *Contrôle industriel de l'étanchéité par traceur hélium*, Edition Société Française du vide, 1992.

NF EN 1330-8, *Essais non destructifs, Terminologie, Termes utilisés en contrôle d'étanchéité*, October 1998.

NF EN 1779, *Essais non destructifs, Contrôle d'étanchéité, Critères de choix de la méthode et de la technique*, December 1999, A1 July 2004.

NF EN 13184, *Essais non destructifs, Contrôle d'étanchéité, Méthode par variation de pression*, November 2001, A1 July 2004.

NF EN 1593 *Essais non destructifs, Contrôle d'étanchéité, Contrôle à la bulle*, November 1999, A1 July 2004.

CHAPTER 12

OTHER TESTING METHODS

This chapter introduces some special methods that meet a particular need or that are of limited or specialized use. As of today, their use requires no specific certification.

12.1 THE ACFM METHOD

The *ACFM* (alternating current field measurement) technique can be used on ferromagnetic materials to test welds, shapes, or surfaces using adapted probes and/or arrays. This method can replace magnetic testing and penetrant testing as it does not require specific surface preparation (paint, surface treatment, etc.). ACFM can also be used in sub-marine testing.

12.1.1 Testing principle

The ACFM technique is based on the absolute measurement of the surface magnetic field generated by an induced electric field that is uniform and parallel to the surface of the part. In the case of an emerging defect, the induced currents are perturbed. The field components Bx and Bz are analyzed when they meet a defect (Fig. 12.1), and their measurement makes it possible to detect the defect's size in length and evaluation in depth.

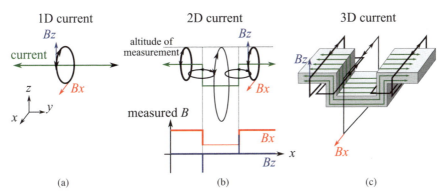

Fig. 12.1 Current (green) and field for a simple conductor (a), amplitudes of field components Bx (red) and Bz (blue), measured perpendicularly to a notch (b), and 3D illustration of currents and fields at the location of the defect (c)

The physical laws involved in the AFCM method are similar to those of magnetic testing and eddy current testing. Though the field is obtained by eddy currents, the disturbances caused by a possible defect are not visualized by colored magnetic powder as in magnetic testing. Similarly, the disturbance of eddy currents is not visualized by its consequences on the resulting field. In AFCM, the field is directly measured, with less dependence on the surface characteristics of the parts.

Defect detection is directional here, as it is in magnetic testing, and it is necessary to operate two perpendicular scans.

The induction fields created by current lines depend on the shapes of these lines; they can be rectilinear or not (rectangular, curved) and homogeneous or not. A notch causes a disturbance in these lines (Fig. 12.2).

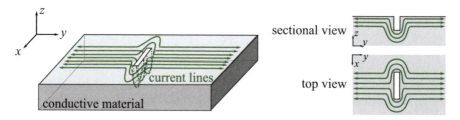

Fig. 12.2 Influence of a notch on current lines

This disturbance (which makes the current lines locally non-homogenous) modifies the shape of the fields at the surface (Fig. 12.3). The illustration shows the conventional representation of the ends (Bz) and of the depth (Bx) of a crack-type defect. The diagram on the right shows a **butterfly** plot corresponding to the combination of the Bx and Bz signals. The advantage of this representation is independent from the probe movement speed. When a defect is detected, data is analyzed to determine the defect sizes without prior calibration.

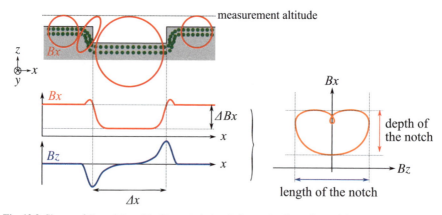

Fig. 12.3 Shapes of Bx and Bz, with ΔBx and Δx in relation to the dimensions of the notch, resulting in a butterfly-like representation

This principle, previously mentioned in the paragraph on magnetic testing method perspectives (cf. § 5.5.7), makes a global dimensioning method for a notch available.

12.1.2 Current and future potential

The ACFM technique was originally developed to search for fatigue cracks in off-shore structures as an alternative to magnetic testing. The advantage was that it was not necessary to clean the marine sediment deposited over time on structures. It was first applied to the legs of oil rigs, to welds on ship's hulls, and to shipyard cranes. Thus, Naval Group (ex-DCNS) divers may use this kind of testing (Fig. 12.4) on an immersed hull in a perfect collaboration between the diver who handles the probe and the inspector who analyzes the signals in real time. Testing before a ship is in dry dock means reducing the time spent in technical stops.

Fig. 12.4 Flexible array probe (C. Laenen, APAVE) and use of ACFM by a diver (Naval Group, ex-DCNS)

This testing technique can be used with or without contact, and without coupling. Testing can be carried out indoors or in enclosed places without endangering the health of operators by the presence of effluents. It is an alternative to magnetic testing and penetrant testing and has the advantage of not requiring prior specific surface preparation (paint, surface treatment, sediments). It can be performed even in the presence of non-conductive paint and coating, up to 5 mm thick.

With the correct probe, it is possible to test all surfaces, even as-welded ones. Fatigue cracks can be rapidly detected on welded structures (cranes, lifting appliances, metallic framework, etc.) or their evolution can be observed with a periodic recording of data.

ACFM can be used to detect, size (in length), and evaluate (in depth) surface cracks in ferromagnetic materials. For other metallic materials (aluminum, stainless steels, etc.) the controlled depth is larger, but the detection sensitivity is then lower, and the depth evaluation is more difficult, requiring a specific calibration.

As a drawback, this technique is less sensitive than eddy current testing, penetrant testing, and magnetic testing. For example, an estimation of the limit of ACFM sensitivity corresponds to the detection of defects 5 mm long and 0.5 mm deep whereas in magnetic testing, this limit corresponds to the possible detection of defects 0.5 mm long and 0.1 mm deep.

12.2 SUPER HIGH FREQUENCY, MICROWAVES, RADAR, AND TERAHERTZ METHODS

The principle of these methods is to emit a beam of electromagnetic waves in the direction of the material to be tested. The beam is then partly reflected by the presence of the material and partly transmitted through it. Comparative measurement of the reflected or transmitted waves with the incident waves offers an insight into the internal characteristics of the tested product or material.

The presence of a defect in a material appears as a local variation in the dielectric properties. By measuring the transmission coefficient and the reflection coefficient, a signal depending on the characteristics (dimensions, nature) of a possible defect is obtained.

12.2.1 Testing principle

Super high frequencies (SHF), ***microwaves***, ***radar waves***, and ***terahertz*** (Fig. 12.5) are the different names of very high frequency electromagnetic waves that can propagate in dielectric materials (electrical insulators) and be partly reflected at the interfaces of media of different natures. These waves are almost entirely reflected by an electrical conductor medium such as a metal.

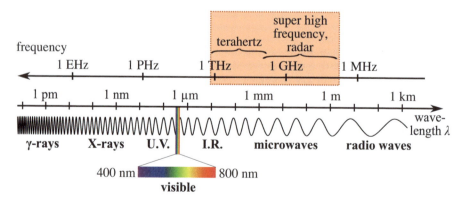

Fig. 12.5 Frequencies usually associated with super high frequencies, microwaves, radar, terahertz methods

The super high frequencies used range from a few hundred megahertz to a few dozen gigahertz, and terahertz frequencies range between a few dozen and a few hundred gigahertz. Materials a few centimeters thick can be tested by the first type of frequencies and the second type for a few millimeters or tenths of millimeter and even hundredths. The detection resolution performs better with terahertz frequencies than with SHF but reversely for attenuation.

Very high frequency electromagnetic waves propagate in insulator materials with a celerity and attenuation that essentially depend on their dielectric permittivity. True dielectric permittivity mainly influences celerity and represents the material's polarization capacity, whereas complex permittivity represents losses that have different origins according to the relevant frequency range.

When a wave is emitted in the direction of a material, it is divided in two at each change of dielectric medium, into a transmitted and reflected wave: either at the surrounding environment at the input interface of the material, the bottom of the material, or the interfaces in the material if a defect is present (Fig. 12.6).

The measurements of the amplitude and reflection (or transmission) coefficient phase represent data on the material and on the possible present defects. Knowing the reflection coefficient of a defect-free material or eliminating it by initial calibration on a "sound" material gives direct access to the defect's characteristics: thickness, overall dimensions, nature, and geometry.

Associating a physical scanning device of the sensor or an electronic one with an array of sensors helps in obtaining an image.

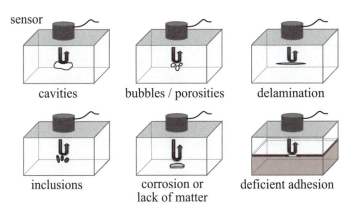

Fig. 12.6 Application of super high frequency methods for the detection of defects in dielectric materials

The equipment for a microwave field is generally made of an integrated generator ensuring the different functions of wave generation (semiconductor oscillator system), analysis and processing of the reflected or transmitted wave, and data or image display. The type of wave can be continuous (composite testing, plastic, etc.) or impulse (concrete in civil engineering structures).

The design of terahertz devices is more complex and mainly uses two technologies:

- One uses infrared pulse generation (measured in femto-seconds), combined with an IR/terahertz converter, which makes spectral analysis possible
- The other is based on a frequency modulation device with a semiconductor directly generating this type of wave

The first one is adapted to surface or small thickness spectroscopy and the second is more convenient for imaging and internal inspection.

12.2.2 Current and future potential

Microwaves and terahertz waves are complementary to other methods that may not satisfactorily meet the requirements of some particular cases. The reason for this lies in certain advantages of these methods:

- Absence of couplant or no particular coupling (propagation in air or vacuum)
- Access required to one face only
- Low power, no ionization, and no intrusion
- Static or dynamic testing
- Measurement takes many parameters into account: amplitude, phase, polarization, frequency, bandwidth
- Wide choice of available probe configurations, adapted to the tested product
- Real-time testing, imaging
- Access to physical properties (microwaves) and to physico-chemical properties (terahertz)

With these types of waves, and particularly terahertz waves, it is possible to analyze the properties of materials in a part of the electromagnetic spectrum that is not often used, giving a means of investigation in the field of frequencies lower than the infrared.

Here are the main applications:
- Detection of vacuum, cavities, porosity rates
- Detection of inclusions, inserts, reinforcements
- Detection and characterization of mold and the presence of water
- Detection and analysis of polymerization and the presence of components, formulation, quantities
- Measurement of levels
- Detection and sizing of cracks and releases
- Measurement of thickness and coating variations
- Characterization of electromagnetic properties

Testable materials can include plastics, composites, rubbers, wood, cardboard, plaster, concrete, insulation multi-materials, etc.

In the 1980s the use of radar NDT grew in civil engineering (GPR technique, ***ground-penetrating radar***) for the physical and geometric characterization of infrastructures such as railway tunnels (detection of a vacuum, for example), buildings, (detection of reinforcement bars, etc.) (Fig. 12.7), and the ground (detection of underground pipes and systems, measurement of pavement thickness, etc.). Radar was first used to track the echo of metallic object like airplanes and ships (metal being considered a total reflector) over long distances, and the US army adapted it to detect buried objects (crashed airplanes).

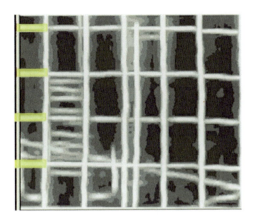

Fig. 12.7 3D radar mapping of a prefabricated prestressed beam (IFSTTAR)

The advantages of this method are high productivity and the absence of contact, the waves being generated by an antenna. Scanned depth ranges from one centimeter to several meters according to the material and to the working wave frequency.

All these examples mostly concern non-metallic materials. With metals, the only application is the detection and characterization of emerging cracks or thickness measurement by comparison, with the advantage of the absence of contact.

The development of these methods, their applications, and their equipment has considerably progressed over the past 20 years. The initial limits linked to technology have tended to diminish thanks to the progress in telecommunication components, miniaturization, the integration of electronic components, and the increasing power of computer calculation. Marketing was slow in the beginning because of extremely high costs, but the situation should improve in the future with the spread of these technologies.

12.3 BARKHAUSEN EFFECT

When a ferromagnetic body is placed in a magnetic field and when excitation is slowly increased, magnetization does not increase in a continuous way but by progressive jumps called "Barkhausen jumps." The variations in magnetization of a ferromagnetic material subjected to a hysteresis cycle are discontinuous and depend on its crystalline structure.

The characterization technique using **Barkhausen noise** shows these sudden (though weak) flux variations that create an electromagnetic wave sensed by a coil. Barkhausen noise is thus a powerful detection tool for structural variations. Its measurement allows a sensitivity to stress aspects and to microstructure, as well. But it is difficult to extract useful information from it.

Measurement setup (Fig. 12.8) includes an excitation circuit, a detection sensor, an acquisition system, and a signal processing module.

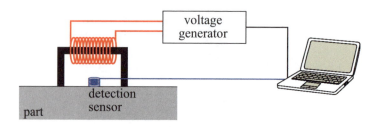

Fig. 12.8 Measurement setup for Barkhausen noise

12.3.1 Testing principle

When there is no external magnetic field, ferromagnetic materials have a spontaneous magnetization that can be microscopically modeled with the introduction of **magnetic domains** or **Weiss domains** (Fig. 12.9), where magnetization is constant (from 0.1 to 100 μm) and separated by **Bloch walls**, transitional layers of variable thickness (0.001 μm approx.). This magnetic microstructure is superimposed to the crystalline microstructure.

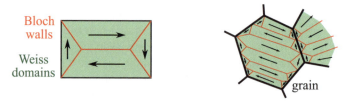

Fig. 12.9 Weiss domains and Bloch walls

When a material is subjected to a hysteresis cycle, the intensity of Barkhausen perturbations for each material is measured, giving the quantity of impurities or of crystal dislocations, and indicating the mechanical properties of this material (Fig. 12.10).

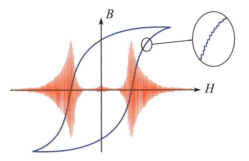

Fig. 12.10 Barkhausen effect (blue) and measured noise (red)

12.3.2 Current and future potential

Some applications focus on the noise level monitoring according to the microstructure obtained (Fig. 12.8) after different thermal treatments (the effective value of the signal is used).

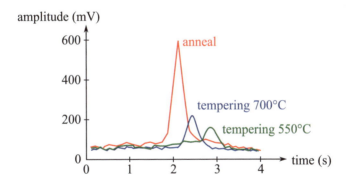

Fig. 12.11 Barkhausen noise according to microstructure

This type of testing can also be applied to stress level measurement. When a magnetic field is applied to a material, it is deformed (magnetostriction), and reciprocally, a deformation caused on purpose promotes certain magnetic directions.

Tensile stress promotes the domains orientated in the direction of the stress, compression promotes perpendicular domains. The shape of the noise signal changes accordingly (Fig. 12.12a).

The purpose of most industrial applications is to characterize the stress state (applied or residual) of cam shafts, pinions, and ball bearings. This way the quality inspection of grinding operation (sound or burnt parts) is completely non-destructive (Fig. 12.12b) as opposed to Nital etching, which is often used.

It can also be used to study the depth reached by surface treatments (cementation, nitriding, etc.) or to sort steel grades, monitor ageing, study decarburization, etc.

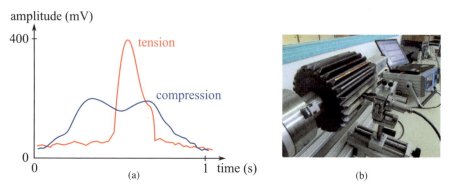

Fig. 12.12 Barkhausen noise versus stress (a), detection of grinding burns (b) (CETIM, for LUFKIN France)

There is an interesting potential in the Barkhausen effect as can be seen in its current use in the measurement of damage on a thin layer in several nanomanufacturing processes (ion etching), but this method is generally too sensitive to other phenomena, and the double aspect "stress – microstructure" often makes interpretation a delicate operation.

12.4 SHEAROGRAPHY

Shearography is used to detect displacements of a few nanometers (a fraction of light wavelength) between two states of an object. It makes it possible to detect various types of damage such as delamination, release, impacts, water intrusion, overheating, cracks, adhesive excess, etc.

12.4.1 Testing principle

Shearography measures ***interference***, which is a way of converting the phase of a wave into an amplitude signal. There is interference when two waves of the same type meet and interact either in a constructive or a destructive way. This requires temporal and spatial coherence conditions, as with a monochromatic plane wave (a laser source in the industry).

The combination of the wave emitted by the laser source (reference) with the wave reflected by the object produces a hologram via the development of a support indicating the absolute displacements. Shearography (or ***speckle pattern interferometry***, or differential interferometry) compares the figure formed by the interference of the waves echoed by the part (rough) surface (***speckle***) with that optically shifted in the plane (sheared) coming from the object (Fig. 12.13). Thus, it is independent from long variations of deformation.

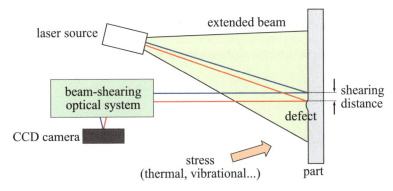

Fig. 12.13 Shearography principle

Shearography is a method based on the combined use of an optical installation generating an image interference and the stressing of the object to be analyzed.

The speckle phenomenon (also called laser granularity) appears when rough surfaces are lit by a spatial and temporal coherent light (a laser). The scattered wave is randomly dephased because of the difference in optical path caused by microroughness (Fig. 12.14).

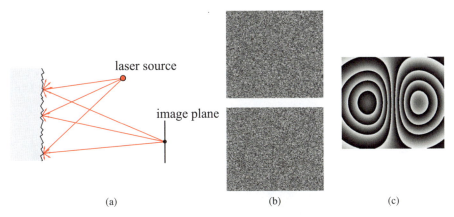

Fig. 12.14 Optical paths (a), speckle patterns without stress and after deformation (b), image of subtracted patterns (c), with a low-pass filtering (M. Viens)

The principle of speckle interferometry is to record the speckle amplitude and phase variation before and after the deformation of an object. A speckle pattern is created by the microscopic roughness of the surface and can be used to study the macroscopic variation with a microscopic resolution. Shearography can show the gradient of out-of-plane displacement, which makes the removal of the general displacements of the inspected part possible.

12.4.2 Current and future potential

Any defect in the integrity of the specimen and causing a surface deformation when a stress is applied will be detected by the comparison of the surface's successive images (Fig. 12.15).

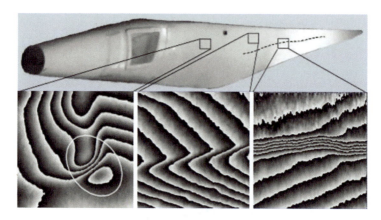

Fig. 12.15 Test of a helicopter blade (M. Viens)

So, when an object is subjected to a stress, the speckle pattern is modified. Subtracting the two speckle patterns (deformed and not deformed) gives a fringe network representing the drift of the surface displacement in a direction normal to the plane.

The size of speckle grains ranges from 5 to 25 μm, which is completely compatible with the resolution of CCD cameras. The devices can be mobile (Fig. 12.16).

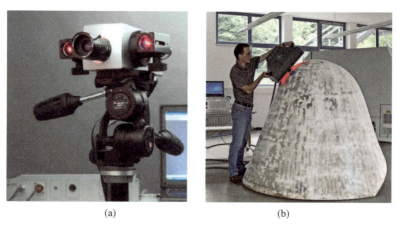

(a)   (b)

Fig. 12.16 Shearography systems (a) (DANTEC DYNAMICS), application to a radome (b) (STEINBICHLER OPTOTECHNIK)

This method performs very well, particularly for focusing visualization on zones where stress is concentrated and analyzing the behavior of structures (vessels) and/or of mechanical assemblies (bonding) when they are subjected to loads.

With shearography, it is also possible to replace strain gauges without contact and to measure displacement and stress in all types of material. Shearography is commonly used to test tires designed for aeronautics and in the retreading of large tires.

The advantages of shearography lie in the way the method is operated:
- Measurement without contact and no surface preparation
- It can be used on any material, of any dimension as long as the part can be subjected to stress
- No consumable is required (films, marking chemical products, etc.)
- Real-time visualization of the whole surface of a part

Disadvantages are essentially connected to laser technology:
- Use of powerful lasers requiring specific safety measures
- The surface has to be highly reflective to scatter incident light, coating is sometimes necessary
- The part or the structure has to be subjected to stress to show possible discontinuities
- The interpretation of the results can sometimes be subjective. Moreover, this method can locate a defect but cannot quantify its dimension, geometry, and depth.

12.5 VIBRATION ANALYSIS, IMPACT-ECHO

Vibration analysis, in its general definition, is the study of the vibrations generated by a machine to detect and identify operational defects such as inadequate tightening, drive shaft misalignment, damaged ball bearing or belts, etc.

The method called *impact-echo* is based on the use of transient stress waves generated by an elastic impact (Fig. 12.17). This technique is mainly used to measure thickness and detect and locate extended defects like cracks and delaminations essentially in concrete structures and masonry works.

A short-duration mechanical impact (produced by a small steel ball or a hammer on the surface of the structure) generates low frequency compression waves (1–60 kHz) which propagate in the structure and are reflected by surfaces (defects, bottom of the part, etc.). A sensor positioned as close as possible to the impact source records the temporal response of the structure.

Analyzing the recorded signal and its spectrum helps find sought after data such as part thickness or the detection and location of a defect.

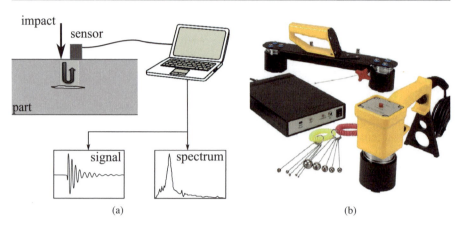

Fig. 12.17 Principle of impact-echo and mobile equipment (HUMBOLDT)

12.6 REFERENCES

ABRAHAM O., POPOVICS J.S., Impact-echo Techniques for Evaluation of Concrete Structures, in *Non Destructive Evaluation of Reinforced Concrete Structures*, Woodhead Publishing Limited, 2010.

CAUSSIGNAC J.M., LE CAM V., ABRAHAM O., DEROBERT X., VILLAIN G., Évaluation et contrôle non destructifs en génie civil, *Techniques de l'Ingénieur*, ref. R1410 V1, December 2013.

CHONG C., *Electromagnetic Testing MFLT/ ECT/ Microwave/RFT Chapter 10 – ACFM Alternating Current Field Measurement*, http://www.academia.edu/, 2015.

COFREND, *Recommandation pour le contrôle des soudures par la mesure du champ d'un courant alternatif - technique A.C.F.M*, Collection LEXITIS Les Cahiers Techniques de la Cofrend, 2012.

DESCOMBES M., JAYET Y., NOEL F., Formation et qualification en contrôle par ACFM: mise en œuvre d'une approche formation/compétence intégrée, COFREND congress, Dunkirk, May 2011.

DESVAUX S., GUALANDRI J., CARREROT H., LAMARE A., Contrôle par effet de bruit Barkhausen: Applications aux roulements à billes et à rouleaux pour l'industrie aéronautique et spatiale, *Instrumentation, Mesure, Métrologie*, Editions Hermès Lavoisier, RS I2M, v4, Nos 1-2/2004: 29–62.

DOVER W.D., DHARMAVASAN S., TOPP D.A., LUGG M.C., "Fitness for Purpose" Using ACFM for Crack Detection and Sizing and FACTS/FADS for Analysis, *Marine Structural Inspection Symposium*, The Society of Naval Architects and Marine Engineers, Arlington, 1991.

IMGC, Méthodes d'auscultation des câbles de précontrainte Impact-écho et tomographie (Recherche de vides d'injection), *Journée Technique Ingénierie de Maintenance en Génie Civil*, September 25, 2015.

INSAVALOR, *Contrôle par ACFM*, Course document, 2013.

KEROUEDAN J., *Conception et réalisation de sondes hyperfréquences pour la détection de micro-fissures de fatigue à la surface des métaux*, Doctorate thesis, Université de Bretagne Occidentale, 2009.

LUGG M.C., *Non Destructive Testing Handbook – Volume 5 – Electromagnetic Testing – Chapter 10: Alternating Current Field Measurement*, Milton KEYNES Editions (UK), 2004.

LAENEN C., LUGG M., Inspection des soudures de fond des réservoirs au moyen de la technique ACFM avec sondes multiéléments, COFREND congress, 2008.

MOUNAIX P., Spectro-imagerie térahertz: Voir autrement, *Techniques de l'Ingénieur*, ref. RE143 V1, May 2010.

MUZET V., BLAIN P., GUILLARD Y., Application de la shearographie à la détection de fissures sur ouvrages d'art, *Bulletin des Laboratoires des Ponts et Chaussées*, 2008: 81–91.

NEYRAT M., *Contribution à l'étude de G.P.R. (Ground Penetrating Radar) multicapteurs: méthodes directes et inverses en temporel*, Doctorate thesis, Université de Limoges, 2009.

PARIZOT F., L'ACFM n'a pas peur de l'eau ni de la rouille, *MESURES*, no. 799, November 2007.

SAGNARD F., REJIBA F., Géoradar: Principes et applications, *Techniques de l'Ingénieur*, ref. TE5228 V1, February 2010.

SALEMI A.H., SADEGHI S.H.H., MOINI R., Thin-skin Analysis Technique for Interaction of Arbitrary-shape Inducer Field with Long Cracks in Ferromagnetic Metals, *NDT&E International 37*, 2004: 471–479.

SMIGIELSKI P., Interférométrie de speckle, *Techniques de l'Ingénieur*, ref. R6331 V1, March 2001.

VIENS M., *Holographie et Shearographie*, Course document MEC761, 2015, Ecole de Technologie Supérieure, Université du Québec.

VILLAIN G., DEROBERT X., ABRAHAM O., CHEKROUN M., COFFEC O., DURAND O., Complémentarité de techniques non destructives pour déterminer les propriétés de différents bétons hydrauliques, COFREND congress, 2008.

ZHANG F., WALASZEK H., GOS C., Application des méthodes électromagnétiques à la caractérisation non destructive des composants mécaniques, COFREND congress, 2014.

CHAPTER 13

NOISE AND HOSTILE ENVIRONMENTS: THE FACTORS OF DEGRADATION OF NON-DESTRUCTIVE MEASUREMENTS

This chapter is dedicated to the study of the factors that may cause the degradation of NDT and/or NDE measurements and cause diagnosis errors. Focus will here be set to the main parameters from which NDT/NDE really "suffers," leaving aside those that can be settled by immediate action, such as with some forms of electronic noise eliminated by simple averaging.

The main degrading factor comes from the material to be tested, as it can be very different from the "perfect" material as defined for each method. Noise coming from the structure will appear in the signal and/or images and degrade information.

Environmental parameters (temperature, etc.) will also be considered and the term "noise" will also be used here in the concept of "hostile environment."

This chapter tries to show the real origins of each type of degradation in order to propose the best solutions. Some of them will be introduced in this chapter; more complex ones will be described in chapter 14.

13.1 STRUCTURAL NOISE

13.1.1 Structural noise and NDT methods

Structural noise is generally introduced as a phenomenon that first degrades the detection operation. Its origin is to be found in the real (micro)structure of the material that the NDT probing wave does not recognize as a homogeneous structure.

This noise is completely deterministic (so it cannot be removed by simple averaging) because (micro)structures are not created randomly but depend on solidification conditions (metallic materials), mix and drying conditions (concrete), fiber installation, and polymerization conditions (composites), etc. Despite it being often too complex to apply the laws of noise generation, a statistical method is sometimes nevertheless used.

In this chapter, a non-exhaustive list of examples comes from certain methods (tomography, eddy currents, etc.) but most of it comes from ultrasonic testing.

13.1.2 Structural noise in ultrasonic testing

Scattering, the origin of noise

Structural noise is directly connected to scattering phenomena. **Scattering** is the deviation of part of incident wave energy by a microstructure. It is caused by discontinuities of acoustic impedance (i.e., heterogeneities in the microstructure elastic properties such as grains, porosities, precipitates, etc.), which are generally called **scatterers**.

As seen in chapter 7, there are three major scattering domains according to the ratio between the mean diameter of scatterers $\bar{d}$ and wavelength λ:

- Rayleigh domain   for $\lambda \gg 2\pi\bar{d}$
- Stochastic domain   for $\lambda \approx 2\pi\bar{d}$
- High frequency domain   for $\lambda \ll 2\pi\bar{d}$.

The concept of scatterer means that size is to be considered cautiously. In the case of metals, depending on the manufacturing process, the distribution of grain sizes around the mean value can be more or less spread. But on the other hand, the size to take into account is the dimension the waves can "see." This may depend on the propagation direction for the same material.

In a heterogeneous medium, any heterogeneity behaves like a multidirectional scatterer (or reflector) and re-emits a part of the energy in all directions. A part of the energy of the incident wave is thus deviated from the initial propagation direction, the other part keeps the direction but is the object of phase-shift and attenuation.

The scattered wave has two contributions:

- A simple scattering contribution. The incident wave is only perturbed once by a single scatterer in the medium before reaching the transducer. This kind of contribution is used by echography and radars because of the equivalence between the time of reception of an echo generated by a scatterer and the distance separating the scatterer from the sensor.
- A multiple scattering contribution. The wave successively interacts with several scatterers before reaching the sensor (Fig. 13.1). This phenomenon is quite significant with scatterers of high diffusion power or those that are highly concentrated (polycrystalline materials, concretes, etc.). The equivalence between the time of reception of an echo and the distance separating a scatterer in the medium from the transducer is no longer valid.

incident
wave

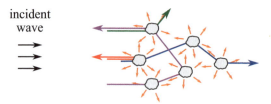

Fig. 13.1 Principle of multiple scattering

Multiple scattering can be neglected when there are not many scatterers, when the impedance jumps between the different phases of the tested material are low, or when the test is considered as being in the Rayleigh domain. In other words, simple scattering is often used as an approximation to simplify the problem.

With polycrystalline metallic materials, most authors use the Rayleigh domain (average size of acoustic heterogeneities clearly smaller than wavelength) in order to develop simple scattering hypotheses for their models. However, assuming multiple scattering is closer to reality.

Scattering is at the origin of attenuation, too

The attenuation that occurs with a probing wave is a common physical phenomenon. It is globally taken into account in all NDT methods, the usual solution being to amplify data on emission and/or reception. Attenuation or more precisely attenuation variations are even the basis for radiographic methods.

But in ultrasonic methods (ultrasound and acoustic emission) the complexity of attenuation of acoustic propagation is a real problem with negative consequences on testing.

Part of attenuation is caused by scattering, because deviated energy is lost for testing. Remember (cf. § 7.1.3) that an ultrasonic wave loses energy when propagating in a real medium via three contributions: absorption, scattering, and beam divergence.

Attenuation by absorption is the result of the transformation of the vibrating mechanical energy into heat. This type of intrinsic attenuation is linked to the viscosity of the inspected material. Attenuation by absorption is considered to be negligible in polycrystalline metals even if some crystal defects like dislocations can contribute.

Attenuation by divergence refers to the contribution of a pure geometric phenomenon related to the divergence of the ultrasonic beam. Energy is diluted in space and consequently decreases on the beam axis. This term can be controlled because it depends on the characteristics of the ultrasonic sensor used (diameter, frequency).

Attenuation by scattering will depend on the domain (Rayleigh, stochastic, high frequencies). If the medium is isotropic, attenuation will be independent from the wave propagation direction. While in an anisotropic medium, such as an austenitic stainless steel weld, it will also depend on the propagation direction.

Ultrasonic structural noise

In presence of scatterers, the attenuation phenomenon is accompanied with a mechanism that creates noise (additional echoes in AScan) because each scatterer makes the energy "explode" in all directions (scattering function).

As a consequence, parts of the wave are deflected in directions that depend on the elastic properties of grains and their geometry and particularly on the orientations of the grain boundaries (especially at high frequencies in the geometric domain). If they do not reach the transducer, they will be lost and the test will be carried out on the remaining, attenuated part. If they reach the transducer, directly or not, they will generate spurious echoes, with different times of flight, leading to structural noise.

By definition, ultrasonic structural noise or backscattered noise (in echo mode) is the part of the ultrasonic wave scattered by the microstructure and received by the transducer. This deterministic structural noise is different from random electronic noise related to the acquisition channel of an ultrasonic signal. On AScan signals, it is represented by the appearance of "spurious" echoes, the amplitudes of which can be significant (Figure 13.2) compared to defect echoes.

The signal-to-noise ratio SNR, comparing the respective amplitudes of the signals of defect and those of noise, can be too low to make a decision about the detection of a possible defect. A SNR higher than 2 (even 3) is usually considered necessary for a diagnosis to be made without ambiguities.

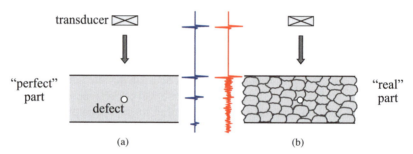

Fig. 13.2 AScan non-noisy signal (a) and noisy, attenuated signal (b)

Distortion of ultrasonic echoes

The beam propagating back to the transducer can appear distorted and also attenuated because the real ultrasonic pulse (polychromatic) is deformed when it goes through the material. This effect is actually different according to the initial spectrum (broadband or narrowband) and grain size, even when in the Rayleigh domain. So, for an emission of the same center frequency, there are two possibilities (Fig. 13.3):

- In the case of a short pulse, with a broadband spectrum, the passage through a material with coarse grains causes a limited decrease in amplitude but also significant frequency filtering: the signal will be highly distorted.
- In the case of a long pulse with a narrowband spectrum, the passage through a material with coarse grains causes a significant decrease in amplitude but a low frequency filtering: the signal distortion will be low.

The idea of intrinsic attenuation

After several studies on reducing attenuation and noise phenomena in order to improve the detection of a defect or amplifying them to characterize material microstructure, it seemed interesting to define a concept of *intrinsic attenuation*, which refers to the energy loss exclusively caused by interactions between the studied medium and the probing wave, which should only depend on the medium (and the wave).

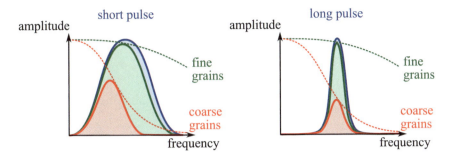

Fig. 13.3 Frequency filtering and signal distortion. Initial broadband spectrum in blue (a) and initial narrowband spectrum (b). Red represents the spectra after the crossing of coarse grains and the green color shows the spectra after the crossing of fine grains.

But this is extremely difficult because the scattered energy does not disappear (contrary to attenuation by absorption) and can be entirely retrieved if an "overall" measurement is considered. The questions are: How can it be defined? By the following definition: "the attenuation refers to the decrease of the ultrasonic amplitude observed with the propagation distance"? On which path? So how can a measurable attenuation value be defined? How can it really be measured, that is, without any influence from the measurement method? How can a reference part be defined? How can it be made?

These questions remain. The LCND (NDT lab) is still trying to work this out, notably with research carried out on an attenuation measurement method based on a point-by-point map of the transmitted beams. The physical realities of the beam (especially its aperture) and the material (its anisotropy) can be taken into account via the decomposition of the beam in plane waves angular spectra and the application of theoretical transmission coefficients in any incidence to each component in plane wave of the incident beam.

13.1.3 Structural noise in eddy current testing

The distribution of the induced eddy currents (EC) in a conductive part complies with laws depending on the magnetic permeability involved. In addition, the penetration depth of EC is much higher (coefficient 10) when the material is non-magnetic.

Thus, local permeability variations, as found in austenoferritic steel testing where the austenitic grains are non-magnetic, will generate "noise" in the acquired signals.

The probe will locally record permeability "jumps" that generate signal variations and therefore noise (Fig. 13.4), which weakens the signal/noise ratio SNR.

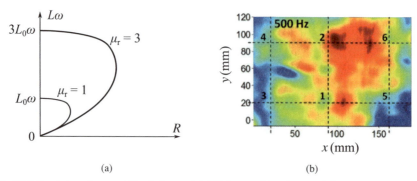

Fig. 13.4 Reminder of permeability influence (a). EC image of 6 reference defects on a stainless steel plate (b) (C. Zorni). The 6 defects are not clearly detected because of important structural noise (red zone).

13.1.4 Structural noise in radiographic testing and tomography

In radiographic testing, scatter can be considered, to a certain extent, as the equivalent of an ultrasonic "structural noise" degrading the dynamics useful for image formation. This is the noise correlated in the image; it is not random and therefore cannot be eliminated by a simple spatial averaging.

The consequences due to scattering appear when the incident spectrum is broadband or when the structures are highly attenuating (large thicknesses, alignment of structures). The number of photons received by the detector is low, image formation is difficult, SNR is low, and defect detection is not ensured. This is also the case when testing is purposely carried out with very low energy. A few experiments rapidly reveal that it is necessary to place the part in a vacuum because air is too highly attenuating!

In tomography, the problem of structural noise or *artifacts* appears instead at low doses because the SNR is often sufficient to compensate these artifacts with the help of simple iterative reconstruction techniques.

The efficiency of a reconstruction method depends on the capacity of the algorithm used to operate in degraded conditions (low number of views, missing data, noise or artifacts, etc.). There are also X photon deflection issues just as for ultrasound in heterogeneous and/or anisotropic media. This kind of issue also appears in phase contrast imaging or in dark-field imaging. The solution consists in reversing and forcing the solution to merge with a probable solution obtained with the help of "a priori" knowledge.

Noise and deflections perturb the tomographic reconstruction as shown in Figure 13.5.

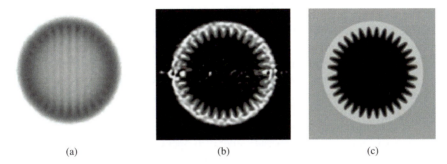

(a) (b) (c)

Fig. 13.5 Reconstitution of the density of a mock-up: radiography (a), reconstructed density (b), original density (c) (INSA LYON)

13.1.5 Structural noise in infrared thermography testing

The materials have to be opaque to the radiation to which the infrared camera is sensitive, in other words, they have to stop the light propagation at their surface. This material property is difficult to appreciate, and several traps may be avoided by the operator, especially with transparent and semi-transparent materials such as glass, plastic, and paint.

In uncontrolled phenomena this could result in "structural noise" such as, for example, the radiation coming from the environment located behind a non-opaque object.

13.2 ENVIRONMENTAL CONSTRAINTS: NOTION OF HOSTILE ENVIRONMENTS IN NDT

The notion of a *hostile environment* is often used in NDT to refer to an environment that will perturb the measure. Three examples are considered: the environmental noise specific to each NDT method used, temperature, and immersion in a liquid.

13.2.1 Environmental noise

Environmental noise is not caused by the material but by the test environment.

In radiography, environmental noise concerns testing carried out in an irradiated environment (for example, radioactive waste container radiographed via an external source), during the autoradiography of an irradiating object (radioactive waste container alone), or possibly when the detector itself is radioactive (scintillator crystals used in PET, positron emission).

For leak testing, one of the benefits of the helium test is that no helium is used for the degassing of tanks and containers (too expensive), so no *background noise* is to be feared.

Experiments in acoustic emission are polluted because of the presence of all the acoustic sources whose frequencies are compatible with the sensors used. Band-pass filtering is often effective in separating the surrounding noise that can sometimes be at high amplitudes, but in different frequency ranges from the searched events.

In magnetic testing, field variations are possible according to the magnetic environment of the part notably because of the presence of masses and/or particular shapes. Pseudo defects are possible.

As described in chapter 10, atmospheric transmission greatly perturbs thermographic measurements. It forces the cameras to operate in three frequency bands for which its absorption will be weaker. But there will always be specific problems with this testing method, such as reflection on glass walls (Fig. 13.6), which adds unwanted rises in local temperatures, for example.

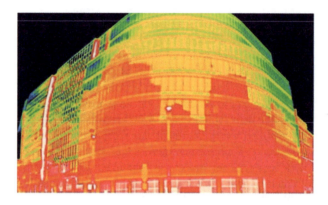

Fig. 13.6 Degradation of infrared thermography diagnosis by reflection on glass surfaces

13.2.2 NDT and high (or low) temperatures

For radiographic testing (or tomography), a hot object has to be insulated (isothermal box), notably because the detectors (including the films) cannot be used at a temperature that differs greatly from the ambient temperature. With these precautions, it is possible to monitor some specific fabrications such as bread baking (Fig. 13.7).

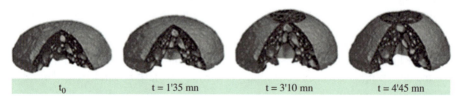

t_0 t = 1'35 mn   t = 3'10 mn   t = 4'45 mn

Fig. 13.7 Tomograms showing the evolution of bread microstructure during the first minutes of baking (P. Babin)

For ultrasound, except for the physical phenomenon that occurs when the piezoelectric ceramics exceeding the Curie point are no longer piezoelectric, the main constraint comes from coupling. Either the measurement is rapid (thickness measurement) and can be performed with a large couplant thickness, or it is slow and the use of acoustic delay lines should be considered to keep the sensor away from the hot zone. Propagation (particularly velocity) depends on temperature, which makes measuring complex, especially if the temperature gradient is unknown.

Acoustic emission testing seems to have the same problem. For example, to obtain a good level of sensitivity, the use of "standard" resonating transducers is recommended. But the physics of these transducers limits them to a contact temperature of approximately 170°C. So waveguides have to be created to make each sensitive element remote and make the contact temperature compatible with the operational limits of transducers (Fig. 13.8).

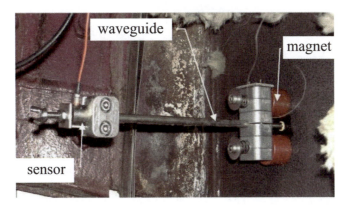

Fig. 13.8 Waveguide on a hot part, in acoustic emission testing (T. Riethmuller)

In penetrant testing, the temperature is limited to 200°C, but conversely the use of conventional products is not recommended below 5°C, because of issues around freezing and the possible malfunction of aerosol sprays.

Temperature also plays a role in magnetic testing when the part approaches the Curie point, which indicates the loss of magnetism. This is already the case for temperatures in the order of 200°C corresponding to pre-heating operations in welding. Otherwise, magnetic testing is carried out with dry indicator products up to 200 or 300°C. Conversely, testing down to –40°C can be performed on oil-based products.

13.2.3 NDT and liquid environments

Magnetic testing can be employed for underwater testing. In this case, an electromagnet is used and is powered by a long wire connecting the magnetization device to a generator outside the water. The lighting used in this technique is of ultraviolet type or white light type and in this case, the indicator product is in red.

Offshore radiographic testing can be carried out without any major technical problems, with the help of a diver – provided divers can protect themselves by moving away from the machine or behind the structures, when depth is not too great. Otherwise, robots are used (Fig. 13.9).

Fig. 13.9 Remotely operated vehicle (ROV) for the inspection of underwater welds (HESCO X-RAY)

Ultrasound has apparently no particular problem in immersion – it is even a specific testing method. But when the fluid is not water but a liquid metal (for example, liquid sodium), wetting difficulties can occur at the liquid-solid interfaces. Gas may appear, the wetting is then composite, and the ultrasound transmission is weakened. This problem may occur on the parts to be inspected as well as on the transducers.

13.3 CONCLUSION

This chapter introduces the origins of signal and NDT image degradation linked to the inspected material. For ultrasonic testing, it has been shown that a precise knowledge of the microstructure ensures that the degradation mechanisms and the solutions that can be implemented are fully understood.

Chapter 14 introduces these solutions in further detail according to the two working axes currently employed: modeling and the analysis of experimental results.

13.4 REFERENCES

ASHBY M.F., *Matériaux, microstructure et mise en œuvre*, Editions Dunod, 1985.

BABIN P., CHIRON H., HOSZOWSKA J., CLOETENS P., PERNOT P., RÉGUERRE A., SALVO L., DENDIEVEL R., DELLA VALLE G., In situ Study of the Fermentation and Baking of Bread Dough by X-ray Tomography, *Publications and Scientific Documentation ESRF Highlights*, 2004.

BUNGE H.J., ESLING C., Texture et anisotropie des matériaux, *Techniques de l'Ingénieur*, M605, 1997.

BREYSSE D., *Non-destructive Assessment of Concrete Structures: Reliability and Limits of Single and Combined Techniques - State-of-the-art Report of the RILEM Technical Committee 207-INR*, Springer, 2012.

CHAIX J.F., GARNIER V., CORNELOUP G., Concrete Damage Evolution Analysis by Backscattered Ultrasonic Waves, *Revue NDT&E* 36, 2003: 461–469.

CHASSIGNOLE B., *Etude de l'influence de la structure m*étallurgique des *soudures en acier inoxydable austénitique sur le Contrôle Non Destructif par ultrasons*, Doctorate thesis, INSA Lyon, 2000.

ESLING C., Texture et anisotropie des matériaux polycristallins – Formation des textures, *Techniques de l'Ingénieur* M3041, 2016.

FEUILLY N., *Etude de l'influence de la microstructure sur la diffusion ultrasonore en vue de l'amélioration du contrôle non destructif des matériaux polycristallins*, Doctorate thesis, Université de la Méditerranée, 2009.

GARNIER V., PIWAKOWSKI B., ABRAHAM O., VILLAIN G., PAYAN C., CHAIX J.F., Acoustical Techniques for Concrete Evaluation: Improvements, Comparisons and Consistencies, *Construction and Building Materials* 43, 2013: 598–613.

GOEBBELS K., *Structure Analysis by Scattering Ultrasonic Radiation*, in R.S. SHARPE, Research Techniques in Nondestructive Testing, chapter 4, New York Academic Press, 1980.

LEGER A., DESCHAMPS M., *Ultrasonic Wave Propagation in Non Homogeneous Media*, Springer, 2009.

LÉTANG J.M., FREUD N., PEIX G., Signal-to-noise Ratio Criterion for the Optimization of Dual-energy Acquisition Using Virtual X-ray Imaging: Application to Glass Wool, *Journal of Electronic Imaging* 13 (3), 2004: 436–449.

MOYSAN J., GUEUDRÉ C., PLOIX M.A., CORNELOUP G., GUY P., EL GUERJOUMA R., CHASSIGNOLE B., Advances in Ultrasonic Testing of Austenitic Stainless Steel Welds. Towards a 3D Description of the Material Including Attenuation and Optimisation by Inversion, in *Ultrasonic Wave Propagation in Non Homogeneous Media*, Editors A. Leger, M. Deschamps, *Springer Proceedings in Physics 1*, Vol. 128, 2009: 15–24.

NAGY P.B., ADLER L., Surface-roughness Induced Attenuation of Reflected and Transmitted Ultrasonic Waves, *Journal of the Acoustical Society of America* 82 (1), 1987: 193–197.

NAYFEH A.H., *Wave Propagation in Layered Anisotropic Media*, North-Holland, 1995.

NEUMANN A.W.E., On the State of the Art of the Inspection of Austenitic Welds with Ultrasound, *International Journal of Pressure Vessels and Piping* 39, 1989: 227–246.

PAPADAKIS E.P., Revised Grain-Scattering Formulas and Tables, *Journal of the Acoustical Society of America* 37 (4), 1965: 703–710.

PLOIX M.A., *Caractérisation ultrasonore des soudures en acier inoxydable austénitique: étude de l'atténuation*, Doctorate thesis, INSA Lyon, 2006.

PLOIX M.A., GUY P., CHASSIGNOLE B., MOYSAN J., CORNELOUP G., EL GUERJOUMA R., Measurement of Ultrasonic Scattering Attenuation in Austenitic Stainless Steel Welds: Realistic Input Data for NDT Numerical Modeling, *Ultrasonics* 54 (7), 2014: 1729–1736.

ROSE J.L., *Ultrasonic Waves in Solid Media*, Cambridge University Press, 1999.

STANKE F.E., KINO G.S., A Unified Theory for Elastic Wave Propagation in Polycrystalline Materials, *Journal of the Acoustical Society of America* 75 (3), 1984: 665–681.

VIDAL F.P., LÉTANG J. M., PEIX G., CLOETENS P., Investigation of Artefact Sources in Synchrotron Microtomography via Virtual X-ray Imaging, *Nuclear Instruments and Methods in Physics Research, Section B: Beam Interactions with Materials and Atoms* 234 (3), 2005: 333–348.

See also the references in chapter 14.

SOLUTIONS TO COMPLEX NDT PROBLEMS: AN OVERVIEW OF DIFFERENT REFLECTION FOCUSES

NDT is an *inverse problem*. An inverse problem consists in finding out the causes of a phenomenon from the experimental observations of its effects. In other words, it reverses the *direct problem,* which deduces the effects from knowledge of the causes. The direct problem (analytic, numerical, empirical, etc.) is related to the principle of causality, which may be more natural for a scientist.

For example, in seismology, locating an earthquake from measurements taken by stations distributed all over the Earth is an inverse problem. And in NDT, finding the position of a defect, its size, etc. from different measurements is an inverse problem, too. Understandably, when the material is heterogeneous and the environmental conditions are difficult, finding the nature and size of the defect can be very complex.

It is like trying to find the color of a drum from its sound... or in the case of NDT, finding the physical properties of a system (considered as a black box that cannot be opened) from the system's responses to known stresses caused by non-destructive processes.

It is impossible to try and solve an inverse problem without thoroughly understanding the direct problem. The general laws with which the problem and its behavior comply have to be understood. When a phenomenon is identified, the most detailed deductive analytic model possible is then constructed.

This chapter will outline two possible working axes, modeling (§ 14.1) and the analysis of results from experiments (§ 14.2-3-4).

14.1 MODELING IN NDT

14.1.1 Direct problem/inverse problem

NDT (or NDE) uses a probing wave with a field that propagates; the associated inverse problems are field inverse problems of two different types: *source inverse problems* and *medium inverse problems*.

In the first case, the medium is presupposed to be known, and source characteristics are searched for (location, extent, amplitude). In the second case, which better represents NDT, the source is presupposed to be known, and the medium characteristics (through which the source-emitted wave propagates) are investigated.

Medium inverse problems can be handled in two ways: with analytical methods and numerical methods.

- **Analytical methods** are generally based on the linearization of the inverse problem by a hypothesis, such as small disturbances of the field by the medium (Born, etc.), an approximation that leads to a simple inversion of the analytic formula close to a direct problem.
- If the intrinsically non-linear character of the inverse problem cannot be avoided by any linearization, iterative **numerical methods** are used (Fig. 14.1). At each step corresponding to a new configuration of the medium, the direct problem is numerically solved, and the simulated results are compared to the experimental data via a cost function in which the evolution is monitored. At each iteration, the cost (the error) obtained is minimized as much as possible. The inversion is supposed to be correct and the process stopped when the error is lower than a limit value defined as the stop criterion of the iterations.

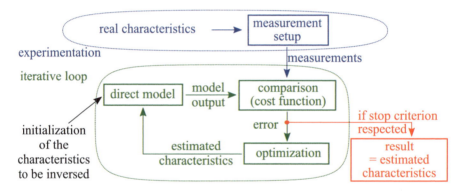

Fig. 14.1 (Iterative) general inversion process

If all the iterative inversion methods are similar to the general diagram, their difference is mostly about the optimization algorithm, which successively modifies the characteristics from the estimate of the error. There are quite a lot of these algorithms: Newton's method and the conjugate gradient method (local methods) and simulated annealing method and genetic algorithms (global methods) are the most commonly used.

The mathematical resolution of inverse problems is difficult because they are **ill-posed problems**: experimental observations only are not sufficient to exactly define all the model parameters, especially for real complex media (in the experimental field) modeled in a direct problem by perfect, homogenized, etc. media. However, resolution of these problems is currently simplified thanks to the incredible improvement in calculator capacities.

It is only in rare specific testing cases (predominantly in the NDE field) that the implementation of a complex inversion is necessary. In usual industrial testing (NDT), where some simplifying hypotheses can be used (isotropy, roughness, temperature, etc.), the direct analytical model is easy to inverse. Sometimes, the parameters are just intuitively adjusted, if necessary.

These analyses can be simplified by the use of signals and/or image processing methods making it possible, a posteriori, to search for the most relevant information from modeled or experimental data. For example, the statistical modeling of a noisy signal does not really help to extract information (defect presence) from the digitized experimental signal. But it gives a better understanding of what is happening, and it can guide the user in the choice of the optimal digital processing.

14.1.2 Analytical models and numerical models

The purpose of analytical methods is to obtain a precise mathematical expression of the solution, whereas numerical methods aim at calculating, with close approximation, the numerical values that express the solution. Analytical modeling is instead limited to simpler phenomena under some precise hypotheses.

For complex cases, numerical modeling tries to find solutions using an approximate approach that is "reliable enough." For example, the *finite element method* uses a meshing of the domain, and the choice of the mesh size is important for the reliability of the predictions.

There are many types of modeling software that are specific to each NDT method and that try to avoid specific situations, but few marketed tools have the potential offered by the software platform CIVA. CIVA is a platform of expertise for NDT (developed by the French Alternative Energies and Atomic Energy Commission, CEA) and is composed of simulation, imaging, and analysis modules that design and optimize testing methods and can predict their performances in realistic testing configurations (Fig. 14.2).

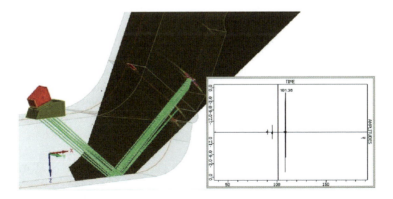

Fig. 14.2 Results of a testing simulation on a complex part and the associated AScan (EXTENDE)

Currently, ultrasound testing (CIVA UT), radiographic testing (CIVA RT), and eddy currents testing (CIVA ET) can be modeled. Complementary modules are available such as guided waves (with the ultrasound module), tomography (with the radiography module), or ATHENA 2D (developed by EDF), a ultrasound finite element module (with the ultrasound module). These modules coexist in the same working environment. To carry out the analysis, its imaging system associated with signals and imaging processing modules (frequency filtering, deconvolution, wavelet analysis, etc.) interprets and assesses the experimental and simulation results.

In ultrasound, the CIVA software offers field and echo calculation modules based on semi-analytic formulas. These modules define the configuration of complex 3D calculations in terms of defects, transducers, and geometries, but their application is limited because of the simplifying hypotheses imposed by semi-analytical methods.

CIVA has been developed and edited by the CEA for more than 15 years with the help of its partners. Today, it is distributed by the EXTENDE Company.

The ATHENA code, developed by EDF R&D, is a code with finite elements based on the resolution of elastodynamic equations expressed in terms of displacement velocity and stresses. It is used in the study of propagation in heterogeneous and/or anisotropic materials and of the interaction of the beam with complex-shape defects.

A hybrid model integrating ATHENA in CIVA has been developed to couple the benefits of semi-analytical methods (rapidity) with finite elements (genericity).

14.1.3 Main industrial stakes in modeling

The role of modeling has now become very important, whether it is in non-destructive testing or characterization, industry, and research. There are several levels of interest but in all cases the reduction of cost and time compared to an experimental process is one of the objectives.

Here are a few examples that are currently being developed by Airbus Group and in the nuclear field where the needs are similar, even though the applications in terms of materials, thickness, and manufactured quantities are different:

- Optimization of the testing processes by defining the appropriate sensor and movement to inspect a specific zone of the part
- Optimization of ultrasonic sensors for the study of specific defects
- Simulation of X tomography to evaluate experimental errors
- Simulations of ultrasonic response to complex defects
- Determination of the mechanical consequences of the defects
- Determination of the material porosity from experimental data
- Understanding of physical phenomena and development of NDT methods by evaluating the influence of parameters, elimination of the shadow zones in sensors
- Support in qualification by showing the performances of an NDT method in extreme cases
- Understanding of the expertise on working sites, assistance in the interpretation of complex results
- Execution of "virtual inspections" at the designing stage of a component

In some domains, modeling is inevitable. This is the case (among others) with ultrasound phased array testing. It is possible to

- Help design methods and probes;
- Generate the delay laws that lead the ultrasonic beam to the desired area;
- Attenuate the secondary lobes by optimization of the size of elements; and
- Anticipate and understand the results from a sometimes complex phased array inspection according to delay laws, part geometry, etc.
- Modeling contributes to the reduction of testing costs by the gain in time obtained, notably by the decrease in the number of experiments to be carried out in order to qualify the methods.

14.1.4 A few examples of modeling in NDT

Modeling ultrasonic structural noise

Because of the multiple scattering phenomena described in the previous chapters, the study of backscattered noise is complex, and there are currently only a few reliable models.

Recent works take into account the function of grain distribution (shape and orientation), as in Voronoi diagrams (Fig. 14.3), in order to describe the microstructure of polycrystalline metals (macroscopically homogenous and isotropic) in a more realistic way than with circles, ellipses, or rectangles. If a Voronoi cell properly represents a grain, the authors give to each one of them an orientation (random for now), that is, a set of elasticity constants from the corresponding, macro-isotropic material.

Predicting the properties of an ultrasonic structural noise is more complex than predicting those of attenuation, and it requires more information. Structural noise comes from the whole volume subjected to ultrasound: it depends on the 3D geometry of the inspected part and on the used source and receiver.

Current research is trying to define the quantities intrinsic to a material that indicate its ability to echo structural noise and that do not depend on part geometries or sensors. These quantities could be used by algorithms, which then would predict the noise in different testing configurations.

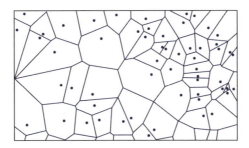

Fig. 14.3 Polycrystalline model using Voronoi diagrams

Modeling of ionizing radiation testing

The parameters necessary for image modeling are those involved during the emission, interaction, and detection of X rays. This kind of modeling is well known. It is also possible, for example, to couple probabilistic methods (Monte-Carlo) in order to correctly model scattered radiation in parts with large thicknesses, using deterministic methods (analytical), which are faster and sufficient for thin and intermediate thicknesses.

Modeling must also take into account the need to reduce received doses while maintaining image quality (IQ). This essential application concerns living beings (the operators and/or the patient in the medical field), where the received dose/IQ compromise is indispensable. But other NDT methods (ultrasound, etc.) follow the same logic, where modeling is used to rapidly find the right form of experimentation.

Thus, modeling in radiographic testing simplifies the adjustment of operative conditions of a radiographic shot without having to make large experiment campaigns, and simplifies the interpretation of radiographic images when explaining the phenomena observed.

Modeling needs also concern specific applications, for example:

- Very high resolution (for living beings and in industry, too). In classic imaging, if a certain dose is needed for a given IQ in a voxel of 1 mm^3, a quantity a billion times bigger is needed to have the same IQ in a voxel of 1 μm^3. The received dose would be huge and so it is necessary to model this irradiation and the future signal (or imaging) processing to better limit possible consequences of deviation.
- The experiment is carried out and based on a single X-ray shot. This is the case when the consequences of the shot physically degrade (heating) the integrity of the material (occurrence of microcracks). Modeling helps to perform this single shot.

Independently from these specific needs, modeling tools are generally efficient and globally, problems are solved. The problem remains, however, for more recent detectors (energy-resolving detectors, for example) or for sophisticated imaging techniques (phase imaging, diffraction imaging, etc.).

Modeling in eddy current testing

The development of testing models using eddy currents is necessary in order to understand electromagnetic phenomena that are particularly difficult to grasp. These types of modeling also quantify the influence of several parameters such as part geometry, lift-off, and the metallurgical characteristics of the inspected parts (and sensors).

Modeling in eddy currents helps to understand and interpret the measurement results, because the signature of a defect can be very complex.

14.1.5 Simulation and statistics: the POD

To determine the soundness condition of a part via NDT, it is necessary to define a certain number of data: inspection method (ultrasound, eddy currents, radiography, etc.), material, considered defects, and implementation (human operator or automated experimentation). Then a decision has to be made on what maintenance to carry out.

But several measurements in an unchanged configuration can lead to different results because of influencing parameters that are difficult to control, such as the real characteristics of the sensors, the material, the inspection conditions, and the human factor. By carrying out several measurement series and making a statistical study that precisely evaluates the method performances, these parameters can be taken into account.

Thus, a *POD curve* (probability of detection) expresses the probability $POD(a)$ of detecting a defect with a characteristic magnitude already known a (such as its size), considering the uncertainties caused by the other influencing parameters previously mentioned.

A POD curve (Fig. 14.4) shows, for example, that the probability of detecting a small defect tends to a null value (0), and the probability of detecting a large defect tends to 1. It also defines a trust band (generally fixed at 95%), which is an approach standardized by the American Defense Department in the 2000s.

Two other notions are sometimes associated with the POD curve:

- The notion of probability of false alarm (PFA): there is no defect in the part but the test detected one. A sound part is removed from production and the problem becomes purely economic.
- The notion of probability of non-detection (PND): there is a defect in the part but the test has not detected it. The error is more serious because safety is at risk.

POD curves quantify the performances of a given inspection method and can be used in the decision process (maintenance). As the essential statistical analyses for the creation of a POD curve are based on long and expensive series of experimental testing, some professionals have suggested shortening the time and reducing the costs of these analyses by coupling the statistic approach and simulation tools. The number of experimental tests is then limited thanks to the implementation of virtual testing (a POD module is also included in the CIVA software).

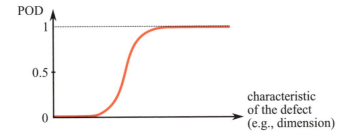

Fig. 14.4 Representation of a classic POD curve in relation to a defect with known characteristics

14.2   DIGITAL PROCESSING OF THE SIGNAL: EXAMPLE OF ULTRASOUND

Signal and image processing is a group of methods (now digital) that in the case of NDT aim to improve the analyses and diagnostic abilities of the operator. Thus, whatever the method used, the non-destructive inspection is carried out with several purposes: detection of the events that are likely to be defects (without adding false alarms), and/or possibly the location, and/or (non-volume or volume) identification and/or sizing.

Much has been written about original signal processing methods bringing improvements to each of these objectives. But globally, there is no universal processing; each situation is unique. This paragraph introduces a few solutions specific to ultrasonic testing, the knowledge of which can help solve new problems.

14.2.1   Prerequisite: consequences of the experimental choices on the analysis

All types of digital processing are based on a digitized signal, of which the number of retained points has to be optimized so that the reconstituted signal enables efficient, but more importantly, valid processing. The *Shannon theorem* handles this operation: the signal is properly sampled if the *sampling frequency* f_s used is higher than twice the maximal frequency f_M included in the spectrum of the signal to digitize: $f_s \geq 2f_M$.

Figure 14.5 shows and introduces the consequences of this choice. To obtain a digitized signal $sd(t)$ identical with the analog signal $s(t)$, its *Fourier transform* $Sd(f)$ is also assumed to be identical, in relation to $S(f)$. As well as the multiplication by the *comb* $cmb(t)$ (digitization effect), the *convolution* by the comb $CMB(f)$ shows that there will be *spectrum overlapping* if the Shannon condition is not respected, in which case the sampling is not performed correctly and all of the resulting processing will be incorrect.

Only maximal frequency is to be taken into account and not the center frequency. Thus, two ultrasonic transducers of identical center frequencies but with different damping (high damping "broadband" or low damping "narrowband") need different sampling conditions.

From the theoretic lower limit defined by Shannon ($2f_M$), an optimal sampling frequency choice has to be correctly made because this value is retained for all subsequent numerical calculations.

For example, Figure 14.6 shows the same ultrasonic signal digitized on $N = 1000$ points at $f_s = 50$ MHz (up) and $f_s = 250$ MHz (down). The calculation algorithm of the Fourier transform (FFT fast Fourier transform) determines that the two amplitude spectra have $N/2$ points each, represented on a basis $[0, f_s/2 = 25$ MHz$]$ for one and $[0, f_s/2 = 125$ MHz$]$ for the other.

The zoomed views of the two spectrums clearly show that the frequency resolution is more accurate for the sampling at 50 MHz, as there is a frequency resolution equal to $f_s/2$ MHz for $N/2$ points, that is, $f_s/N = 50$ kHz. Whereas for the other it is 250 kHz.

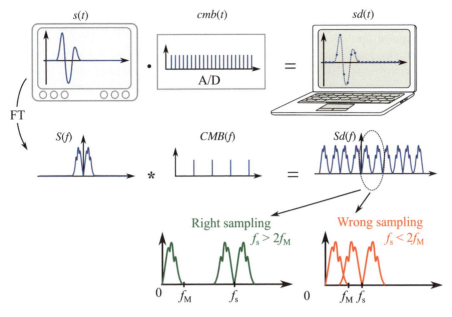

Fig. 14.6 Illustration of the Shannon theorem

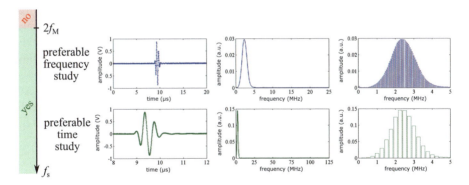

Fig. 14.5 Consequences of the choice of sampling frequency

As a consequence in ultrasound, it is preferable to carry out a low oversampling for a frequency study and a high oversampling for a time study. When there is uncertainty, it is possible to digitize twice (if the signal is not transient, of course) or highly oversample when digitizing and re-sample (by removing one point out of two, for example) before carrying out frequency processing.

14.2.2 Improvements in the detection (and location) of defects

By temporal, spatial, and frequency averaging

Temporal averaging of N signals obtained at different moments at the same position helps improve the SNR by a ratio $\sqrt{N}$ when the noise is random. In most cases, the random component related to electronic noise is low compared to the noise generated by the grains. Thus temporal averaging turns out to have low efficient on ultrasonic signals.

Spatial averaging uses several signals obtained from close positions. Basically, the idea is to make the responses "random" if there are only grains and "deterministic" if there is a defect. The idea is interesting, but it almost requires having found the defect to improve the SNR in order to better detect the defect.

Frequency averaging, or split spectrum processing (SSP), is based on the hypothesis that the laws of amplitude variation according to frequency differ from what the grains and the defects echo. The basic principle consists in obtaining several signals using different frequency bands and recombining the results.

By correlation of signals

Correlation consists in measuring the link between two events. The function of correlation is a function of similarity in shape and time between two signals $s1(t)$ and $s2(t)$. The function is maximal (Fig. 14.7) when the signal $s2(t)$, delayed by a time T, has a shape as close as possible to the signal $s1(t)$ (maximum of similarity).

Cross-correlation, carried out between a measured signal and a reference signal, is different from *autocorrelation*, which is made on the signal itself. Cross-correlation helps to find the position of the reference event in the measured signal. It can be applied in the location of defect echoes lost in noise or in measurements of transit time of ultrasonic waves, even when the signal is distorted.

In the latter case, the high performance of the cross-correlation is to be noticed, as the value obtained (maximal) does not depend on the operator, contrary to techniques using chronometry, where the operator chooses times at an extremum, at the crossing of a threshold, or at the zero crossing, etc.

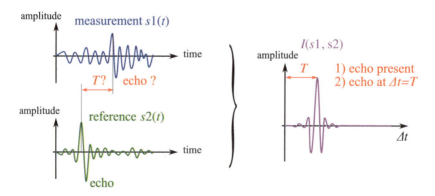

Fig. 14.7 Principle of cross-correlation; by shifting $s2(t)$ of T, the shape $s1(t)$ is obtained

The autocorrelation, basically, searches in the signal for the place where it is the most similar to itself. That is, by searching for periodicities inside the signal, applications for this type of correlation can be found in thickness measurement (periodicity of multiple roundtrips).

By deconvolution of signals

A transducer does not provide a perfect "peak" for each impedance discontinuity that has been met. Its transfer function (damper, etc., see § 7.2.7) transforms the wave representation, extends it, or even gives an overlap of the echoes provided by each impedance discontinuity present in the material, resulting in a loss of temporal resolution. The mathematical description of the problem is expressed by the following formula:

$$s(t) = e(t) * h(t) + b(t)$$

where $s(t)$ is the signal received by the transducer, $h(t)$ is its pulse response, $e(t)$ the reflectivity of the medium, $b(t)$ the noise, and the **convolution** operation is symbolized by an asterisk.

Improvement in this resolution can be obtained by a frequency increase, but this possibility is quickly limited by the increase in ultrasonic attenuation. Another solution consists in implementing a **deconvolution**, or restoration of the reflectivity, which amounts to eliminating the transfer function of the sensor and so restoring the "peaks" of the material reflection.

By working in the Fourier domain and by neglecting the noise, the reflectivity transform of the medium is obtained with a simple division. But this frequency deconvolution is only valid for transfer function values (denominator) that do not vanish. That is why some people work on temporal deconvolution (reversing the convolution function); unfortunately this is longer and more complex because it is based on a priori information about the evolution of the pulse response during the propagation.

By time-frequency analysis

Analysis by Fourier transform is even more accurate, as the observation of the signal is very long (even infinite according to Fourier). Actually, it is by observing two sinusoids of very close frequencies for a very long time that a difference can be pointed out. As a consequence, it is not possible to "locate" in time the frequencies found. This is the challenge met by the **time-frequency** methods: by replacing the infinite sinusoid of the analysis with the appropriate **wavelet** to obtain both the frequency information and its temporal location.

As the temporal and frequency domains taken alone are not individually convincing, the use of time-frequency analysis is justified when searching for an indicator. The LCND laboratory worked on alkali-reaction damage on concrete and showed that time-frequency analysis can differentiate sound concrete from damaged concrete (Fig. 14.8) during the setting and hardening of concrete (over 80 days), whereas the obtained ultrasonic signals and their spectra were not easy to distinguish.

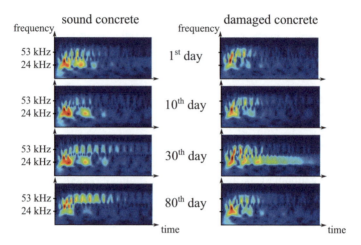

Fig. 14.8 Time-frequency analysis of a few AScan signals acquired during the setting of sound concrete and concrete damaged by alkali-reaction (LCND)

14.2.3 Improvements in the characterization of defects

The techniques of defect characterization mainly focus on the geometric form of the defects in order to classify the defects into two categories: "non-volume" (crack type) or "volume" defects (inclusion or blowhole type).

Some experts search for the shape of the *echodynamic curve* obtained by keeping the received maximal amplitudes when moving with respect to the defect. Others choose to calculate shape criteria such as rise time, echo duration, and fall time to try to statistically differentiate crack- or inclusion-type defects. A lot of parameters can be used, and certain systems are able to handle about a hundred of them. Some established libraries of reference transfer functions in order to compare them with those of the measured obstacle, etc.

With the help of neural network methods, some experts are trying to define the optimal relations between these parameters. The expert systems (the higher or final step of classification) gather processing, make classifications, and arrive at a decision according to these data.

But one of the major problems of these approaches is that they are too dependent on the "transducer - measuring chain - part" system implemented and on the need for learning, which requires a lot of typical data.

As a conclusion, though several of these methods have been studied and validated case by case, none of them is really used in industrial testing today.

14.2.4 Improvements in defect sizing

The standard application is the sizing of a crack by measuring the precise time of flight corresponding to the echoes diffracted by the two crack ends after displacing the transducer. The exact measurement of the transit time of an ultrasonic wave is not obvious, because it depends on the degradation of the echo that carries the information. According to the authors, many different forms of time measurement can be considered:

- When passing a pre-defined threshold (typical case of thickness measurers)
- At the HF signal maximum
- At the maximum of the module of the associated analytic signal (the signal envelope)
- At the bottom of the same envelope or when passing a determined threshold
- At the maximum of the cross-correlation function obtained with a reference signal
- At the maximum of the autocorrelation function, if a periodicity exists

The width of a small defect can be determined with the help of a frequency analysis of the signal diffracted at the same time by the two ends of the defect for a certain position of the transducer. There is a maximum (point) in the spectrum where the two waves are in-phase, that is, the time of flight difference corresponds to a multiple of the wavelength.

14.3 DIGITAL PROCESSING OF NDT IMAGES

Generally in image analysis, the purpose is to improve the quality of the information (for example, there is a defect or there isn't), while reducing the quantity of information that is maximal during the acquisition.

The diagram below (Fig. 14.9) introduces the domains of action of image processing; when information quantity is maximal (acquisition of mega octets), quality is often minimal (in NDT, is there a defect?). And, after different types of processing, defect detection needs only one bit (0–1), but this information is very rich…

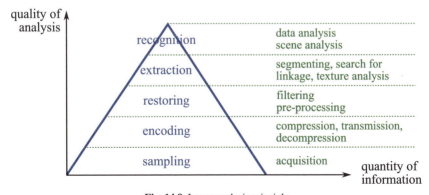

Fig. 14.9 Image analysis principle

14.3.1 Prerequisite: consequences of experimental choices on the analysis

For radiographic images, eddy currents, etc., the resolutions of digitization used concern only the amplitude and spatial sampling. The ultrasonic image is very different because it also integrates temporal sampling. For 3D ultrasonic data volume, 4 pitches of digitization are handled (Fig. 14.10).

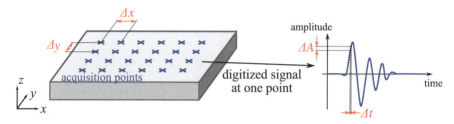

Fig. 14.10 The four digitization pitches for a 3D ultrasonic volume

Shannon's criterion has to be respected in all cases. The sampling frequency has to be higher than twice the maximal frequency of the image. This frequency is expressed in Hz when it comes from time or in m^{-1} when it is a spatial frequency. For the scale of amplitudes, we talk about coding dynamics, expressed in bits.

14.3.2 Improvements to the image: pre-processing

Image quality refers to good visual rendering. It depends on the way colors or grey levels are distributed. Two notions define the image:

- Its *dynamics* – the difference between maximum and minimum effective coding of the image
- Its *contrast* – the difference between the levels of the objects and the image background

Contrast is linked to problems of signal-to-noise ratio but is largely insufficient to describe the quality of an image, because contrast is qualitative. The object in the image has to be defined and identified prior. There is no standard rule, which is explained by the variety of images.

Many image improvements are in fact 1D signal processings applied to each different signal constituting the image or extended to 2D data. Thus, the technique of deconvolution is applied to 2D; the reference signal can be replaced by a group of reference signals, which constitutes a 2D signature of an object. The temporal, spatial, or frequency averaging techniques (such as split spectrum processing) are often used in signal processing to improve the SNR of an image.

The image offers another possibility: matrix filtering. If the information in the image is considered stable and noisy with a noise of a random nature and Gaussian type, it is possible to calculate an average over the neighbors of a pixel. This means inserting a window into the image and giving the average value to the center pixel. To eliminate interference, the median value is instead used.

The color distribution in the image can be analyzed with a histogram. A linear reframing can restore a complete dynamic to the image. The *flattening* operation, where the histogram is extended, can improve the contrast of objects in an image when it is not homogeneous.

This is the case for radiographic images that have large variations in grey levels, mainly because of the different thicknesses that are filtered through, often showing an image background gradient. Removing this gradient enables an improved visualization of low contrasting objects or small dimension objects.

14.3.3 Image processing: information research

The *2D Fourier transform* plays the same role as in signal analysis: it shows another image representation by giving information about spatial frequencies. It can be used for a low frequency filtering of the image and to operate a flattening. It can also be used to determine the orientation of objects in the image.

Segmentation consists in identifying the different objects of the image and finding out their positions. It is the final image processing result before making a decision, but sometimes this segmentation operation is not complete. This is the case, for example, when the surface percentage of inclusions in a radiographic image is still sufficient even though the quantity or how they are spread on the image are unknown.

Segmentation by thresholding
The *boundary* of an object is the transition zone with the image background. An ideal image has boundaries without thickness, but the acquisition systems create quite blurred boundaries either because of the noise or because of the intrinsic blur of the acquisition system.

There are two classic solutions for image *thresholding*: the first is used with a priori knowledge obtained from a reference image. If the image is more complex, the second solution consists in using a histogram and fixing thresholds in the valleys.

Segmentation by contour research
Contour is a notion similar to that of boundary. By definition, a contour refers to a 1-pixel thickness line and is found inside the boundary zone between the object and the image background. The contour, with a known direction, can be defined by adding a 1D filter of type [−1 1], or by using 2D matrix filtering (north, south, etc.). If the searched contours have no particular direction, Laplacian or Sobel filters are used.

Segmentation by texture analysis
The *texture of an image* is the way the representation levels are distributed. Images of natural texture are sand, sea, grass, etc. The purpose of texture analysis is the identification of objects based not on their only representation level but on criteria of level distribution within the object. One of the fundamental tools of texture analysis is the *co-occurrence matrix* or *2-dimensional histogram*. The distribution of levels between two points separated by a vector, for which distance and orientation have to be defined, is studied. Thus, for the image of an object on a background, information about the object, the background, and the object-background border are obtained.

14.4 NDT DATA FUSION

Data fusion is the most recent development in imaging techniques. It combines several different modalities of NDT (XR, US, EC, etc.) from confirmed information to come to a decision.

Data fusion may be "simple" image resetting, as in industrial inspection (Fig. 14.11) or the medical field where merging anatomic and functional imaging is often required. It enables, for example, a tumor detected in the image to be located in the body.

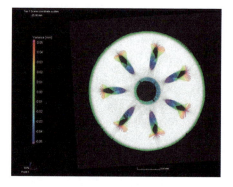

Fig. 14.11 Calibration of a tomographic system by adjustment with a CAD model (NIKON TECHNOLOGIES)

Another classic industrial application consists in combining two techniques: X-ray testing (high accuracy for details in the x, y plane) and ultrasound testing (for depth location z). To improve the contrast, merging images of the same modality can be considered. In particular, the fusion of radiographic images obtained with different intensities can solve the problem of the inspection of parts with different thicknesses.

Data fusion can also be more complex. The LCND lab showed that data fusion based on fuzzy logic and the possibility theory can produce solutions for the characterization of concrete. After numerous experiments on various samples presenting different values of the studied parameters (creation of a database to extract empirical laws), each measurement of observable quantity, such as the velocity of ultrasonic waves, resistivity, radar amplitude, etc., is likely to provide an assessment of the parameters (or indicators) that are searched for, such as porosity, water saturation, etc.

This assessment is more or less accurate and confidence can be variable according to the reliability and sensitivity of observable quantity for each parameter. Data fusion consists in gathering all the assessment from all techniques (each one is associated with a reliability, sensitivity, and uncertainty measurement), and making a compromise out of them in order to choose the most plausible answer (Fig. 14.12).

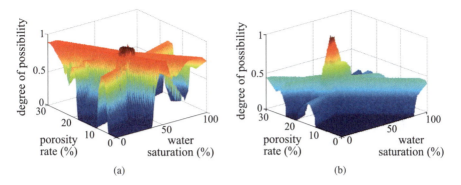

Fig. 14.12 Distribution of fused possibilities showing a delicate (a) and reliable (b) decision

Generally, data fusion tries to reproduce human logic but using a greater amount of data. So, when the initial data converge towards the same assessment of indicators, the result is close to this assessment and the final reliability is reinforced. However, when they disagree, the chosen operator of the combination has to correctly handle the conflicts (for example, by discarding a source of information that disagrees with all the others, which is then considered as having failed, etc.) and must give the most plausible assessment according to all the reliabilities, sensitivities, and inaccuracies.

Data fusion helps to establish a diagnosis. The example of a doctor, who takes into account information based on the observation of a patient, his or her answers, and the several examinations carried out, before synthesizing the information obtained and making the diagnosis, is quite judicious.

14.5 POTENTIAL: RC-CND, RE-CND, SHM

Numerous industries and laboratories are interested in non-destructive testing. They face classic problems where some hypotheses can simplify testing operations and obtain optimal or even perfect situations, but also complex problems, where very few simplifying hypotheses can be validated and where testing is more difficult and some-times almost impossible to carry out.

The first case focuses mainly on industrial testing. It can be carried out by level 1, 2, or 3 certified operators, working in service provider companies in the majority of cases, because it corresponds to an industrial logic of rapidity and quality where processes have to be simple and robust. The second case, less frequent, globally corresponds to R&D needs. This chapter introduces the benefits of potential tools in modeling and experimentation but also the fact that processing is not universal, it is specific to a particular case.

Within the framework of constant improvement in the general quality of instal-lations, the LCND lab studied a particular idea. Over a few years, the concept of *RC-CND* (*Recommandations de Conception des structures basées sur les règles de CND* – Recommendations for Structure Design based on NDT Rules) has been developed. The purpose is to help a structure designer make the best decisions in terms of dimensions and accessibility prior to production. These decisions have to take into account all the NDT techniques (X-ray testing, ultrasound testing, etc.) that will be carried out in manufacturing but also in operation during the life cycle of the components. That way, the testability of a structure will at least be made possible and testing will become safer, faster, and less expensive.

The LCND lab worked on two different approaches for communicating the necessary information to designers with no knowledge about NDT (this was the initial hypothesis).

The first approach consisted in implementing an exhaustive database (expert system) that could propose design "quantitative" verdicts to the designer in order to make future testing easier.

The principle is that each NDT method is initialized as "perfectly adapted" and is given a maximum grade. Then, for each case of precise design, each method is then more or less downgraded according to the material, the types of potential defects (manufacturing or operational defects), and the sought after characteristics (detection, location, identification, and dimension). If the final grade is sufficient, the design is considered adapted to future NDT. If the final grade is too low, the design has to be modified.

Because of the complexity resulting from dealing with so many input-outputs, and the "black box" nature of the result, the chosen solution consisted in suggesting simple "qualitative" rules and recommendations. Thus, after the reading and analysis of French RCC-MRx and RSE-M (a set of technical rules for designing, building, and monitoring nuclear installations), we identified the situations likely to cause a conflict between design and NDT, analyzed them, and wrote recommendations based on NDT rules in understandable words for a non-expert.

On the same principle, the LCND lab has recently started a study integrating the concept of *RE-CND* (*Recommandations d'Elaboration des matériaux et structures basées sur les règles de CND* – Recommendations for the Elaboration of Material and Structures based on NDT Rules). This time, the purpose is to guide steelworkers, founders, blacksmiths, and welders in order to make the best decisions before production, taking into account all the NDT tests to come (ultrasound testing, X-ray testing, etc.). In the same way as for RC-CND, the testability of the material will at least be made possible, and testing becomes safer, faster, and less expensive.

Here is an example of an application concerning welding. In this case, we are working in the framework of both RC-CND and RE-CND, as a weld is both a (welded) structure and a metallurgical elaboration. The LCND lab has developed with EDF R&D a model called *MINA* (modeling an isotropy from notebook of arc welding), which predicts the orientation of grains (Fig. 14.13) in a multi-pass austenitic stainless steel weld made with the SMAW technique in order to predict ultrasonic propagation.

This grain orientation, that is, the orientation of local elastic properties, is obtained from information written in the welding notebook, from parameters specific to the welding process and from rules resulting from crystalline growth. The sequence order of the passes is a particular parameter of order 1 according to ultrasonic propagation. This MINA model is an important step forward in the field of NDT predicting.

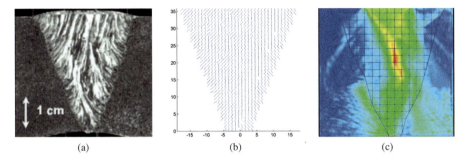

Fig. 14.13 (Destructive) macrograph of a weld (a), non-destructive prediction via MINA (b), making possible the simulation of ultrasonic propagation via ATHENA (c)

It is more relevant than statistic predicting models for orientations in welding, and it avoids the errors observed in the use of a model that is more or less representative of real complex welding.

The use of MINA is also considered when developing a component in order to predict the best welding conditions (order of passes notably), thus simplifying the ultrasonic testing to come in terms of feasibility, economy, and performance.

Another concept is currently being developed: **SHM (*structural health monitoring*)**, or integrated structure testing systems. These systems are often based on non-destructive technologies (ultrasound, eddy currents, etc.) but also on vibration analysis, stress measurements, etc. The principle is real time monitoring (or at regular intervals) of the structure's integrity in which the testing system (material and data processing) is integrated and autonomous.

It detects (and locates) a defect as soon as it appears. The main interest of SHM is then to simplify maintenance operations of a structure through the integration and automation of the entire testing process. They are often referred to as *smart structures* (or "smart material").

The main difficulty with these systems is the integration of components (sensor, energy source, wires, data processing system, etc.).

Currently, an important amount of R&D work is focusing on the development of wireless autonomous sensors, the study of their lifetimes, and their miniaturization. The suggested solutions generally integrate sensors network (piezoelectric, optic fibers, etc.) with the associated data processing systems.

A report of July 14, 2015 written by the MRRSE (Market Research Reports Search Engine) predicts an international market for SHM of approximately USD 1.89 billion in 2020, which means 24.7% growth between 2015 and 2020. Potential savings are considered highly important for operating costs, and conditional maintenance will be simplified, notably for civil engineering (bridges, dams), aeronautics (airplanes), and energy (wind power).

This economic perspective confirms all the scientific potential introduced in this book, for each method, and clearly shows that NDT still has a promising future ahead.

14.6 REFERENCES

ACHENBACH J.D., Quantitative Nondestructive Evaluation, *International Journal of Solids and Structures* 37, 2000: 13–27.

APFEL A., *Etude de l'influence des conditions de soudage sur les possibilités de contrôle ultrasonore d'une soudure. Caractérisation non destructive et in situ d'une soudure austénitique multipasses*, Doctorate thesis, Université de la Méditerranée, 2005.

BOUTEYRE J., Les CND d'EADS ASTRIUM au service du monde naval, colloque Naval Meetings 2013, Bordeaux, November 13–14, 2013.

BEZDEK J.C., DUBOIS D., PRADE H., *Fuzzy Sets in Approximate Reasoning and Information Systems*, Kluwer Academic Publishers, 1999.

BLOCH I., *Fusion d'informations en traitement du signal et des images*, Hermes – Lavoisier, 2003.

BOUCHON-MEUNIER B., MARSALA C., *Logique Floue, Principes, Aide à la Décision*, Hermes – Lavoisier, 2003.

BOUVARD D., CHAIX J.-M., DENDIEVEL R., FAZEKAS A., LÉTANG J. M., PEIX G., QUENARD D., Characterization and Simulation of Microstructure and Properties of EPS Lightweight Concrete, *Cement and Concrete Research* 37, 2007: 1666–1673.

BREYSSE D., LARGET M., SBARTAI Z.M., PLOIX M.A., GARNIER V., Combiner et fusionner les mesures de CND pour mieux estimer les propriétés du béton, *European Journal of Environmental and Civil Engineering* 13 (4), 2009: 501–515.

CORNELOUP G., MOYSAN J., MAGNIN I.E., Ultrasonic Image Data Processing for the Automatic Detection of Defects, *Ultrasonics* 32 (5), 1994: 367–374.

CORNELOUP G., MOYSAN J., MAGNIN I.E., BScan Image Segmentation by Thresholding Using Cooccurrence Matrix Analysis, *Pattern Recognition* 29 (2), 1996: 281–296.

CORNELOUP G., GARNIER V., Analyse ultrasonore du temps de prise de béton compacté roulé, *Materials and Structures* 29, 1996: 295–304.

CORNELOUP G., MOYSAN J., GARNIER V., GUEUDRE C., Image Analysis Dedicated to the Nondestructive Examination of Structural Materials, 14th World Conference on NDT, New Delhi, India, Vol. 3, December 1996: 1785–1788.

DEPARTMENT OF DEFENSE – HANDBOOK, *Nondestructive Evaluation System Reliability Assessment*, MIL-HDBK-1823A 2009, SUPERSEDING MIL-HDBK-1823, 2014.

DORVAL V., *Modélisation de la propagation ultrasonore dans une structure métallurgique diffusante. Application au CND*, Doctorate thesis, Université de la Méditerranée, 2009.

FREUD N., LÉTANG J.-M., BABOT D., A Hybrid Approach to Simulate Multiple Photon Scattering in X-ray Imaging, *Nucl. Instr. Meth. B* 227, 2005: 551–558.

GEORGIOU G.A., *Probability of Detection (PoD) Curves, Derivation, Applications and Limitations*, Research report 454, HSE Books, 2006.

GROS X.E., *NDT Data Fusion*, Arnold, 1997.

GUEUDRÉ C., LE MARREC L., MOYSAN J., CHASSIGNOLE B., Direct Model Optimisation for Data Inversion. Application to Ultrasonic Characterization of Heterogeneous Welds, *NDT&E International* 42, 2009: 47–55.

GUEUDRÉ C., MOYSAN J., CORNELOUP G., Radioscopic Images Segmentation by "Edge and Area" Combined Approach for Weld Geometric Characterization, *Research in Nondestructive Evaluation* 12, 2000: 179–189.

GUEUDRÉ C., MOYSAN J., MERCHI F., HENAULT E., CORNELOUP G., Traitement numérique d'images ultrasonores de haute fréquence pour la caractérisation inclusionnaire d'aciers à roulements, *Matériaux et Techniques* nos. 3-4, 2003: 35–40.

JING YE, MOYSAN J., SONG S.J., KIM H.J., CHASSIGNOLE B, GUEUDRÉ C., DUPOND O., Influence of Welding Passes on Grain Orientation – The Example of a Multi-pass V-weld, *International Journal of Pressure Vessels and Piping* 93-94, 2012: 17–21.

LEMOINE T., Imagerie médicale par rayons X, Traitements d'image 3D, *Technique de l'Ingénieur*, June 10, 2015.

MASSACRET N., MOYSAN J., PLOIX M. A., JEANNOT J.P., CORNELOUP G., Modeling of Ultrasonic Propagation in Turbulent Liquid Sodium with Temperature Gradient, *Journal of Applied Physics* 115 (20), 2014.

MOYSAN J., DUROCHER A., GUEUDRE C., CORNELOUP G., Improvement of the Non-Destructive Evaluation of Plasma Facing Components by Data Combination of Infrared Thermal Images, *NDT&E International* 40, 2007: 478–485.

MOYSAN J., CORNELOUP G., SOLLIER T., Adapting an Ultrasonic Image Threshold Method to Eddy Current Images and Defining a Validation Domain of the Thresholding Method, *NDT&E International* 32 (2), 1999: 79–84.

MOYSAN J., CORNELOUP G., SOLLIER T., A General Thresholding Method for NDT Images: the Case of Eddy Current and Ultrasounds, *Materials Evaluation* 59 (8), 2001: 955–960.

NAGASO M., MOYSAN J., BENJEDDOU S., MASSACRET N., PLOIX M.A., KOMATITSCH D., LHUILLIER C., Ultrasonic Thermometry Simulation in a Random Fluctuating Medium: Evidence of the Acoustic Signature of a One-percent Temperature Difference, *Ultrasonics* 68, 2016: 61–70.

PAYAN C., ULRICH T.J., LE BAS P.Y., SALEH T.A., GUIMARAES M., Quantitative Linear and Non-linear Resonance Inspection Techniques and Analysis for Material Characterization: Application to Concrete Thermal Damage, *Journal of the Acoustical Society of America* 136 (2), 2014: 537–546.

PLOIX M.A., GARNIER V., BREYSSE D., MOYSAN J., NDE Data Fusion to Improve the Evaluation of Concrete Structures, *NDT&E International* 44 (5), 2011: 442–448.

PLOIX M.A., GUY P., EL GUERJOUMA R., MOYSAN J., CORNELOUP G., CHASSIGNOLE B., Attenuation Assessment for NDT of Austenitic Stainless Steel Welds, 9th European Conference on NDT (ECNDT), Berlin, September 25–29, 2006.

SAINTHUILE T., *Récupération d'Énergie Vibratoire pour Systèmes de Contrôle Santé Intégré de Structures Aéronautiques*, Doctorate thesis, Université de Valenciennes, 2012.

See also the references in chapter 13.

QUALIFICATION AND CERTIFICATION

A.1 INTRODUCTION

The activities of non-destructive testing implemented in an industrial context are associated with a training scheme including qualification and certification of personnel, providing and ensuring that they have the knowledge, abilities, and experience required to carry out the tests in accordance with standard practices.

This scheme is available in different modalities according to the relevant industrial sector or to the specificities of a particular manufacturing method or use. There were several reasons that led the users of non-destructive testing methods, as well as manufacturers, to set up this type of scheme as a guarantee for the competence level of the operators:

- Consequences in matters of individual safety, environment, or for the economy are generally so important that the probability for inspected parts or structures having undetected defects must be as low as possible.
- The human factor in the implementation and interpretation of a test requires an expertise relying on several years of practice.
- The implementation of the methods is closely linked to products in the industrial sector, thus experience and knowledge of the companies and their products are imperative.
- The qualification of testing operators must not be impaired by a lack of practice, it has to be checked periodically before being validated again.
- Not only knowledge has to be validated, but abilities in real situations too, in an industrial context, and in the respect of standards.

This scheme is a certification system for individuals in which the different actors have a stake. It ensures a strong adequacy between the available competencies and the needs of companies.

A.2 HISTORY

The certification of personnel in charge of non-destructive testing was first developed in the United States in the 1950s and 60s, and then in France in the late 1970s. The requirements in quality assurance of nuclear plants notably led to the set-up of schemes guaranteeing that the personnel executing the inspections had the necessary level of qualification. From this time on, other industries followed (aeronautics). Nowadays many companies from different sectors certify their personnel or call on the services of certified providers in the context of a client-supplier relationship or to meet regulatory requirements.

In 1973, at the international conference of NDT in Warsaw, several countries considered a joint position on training, qualification, and certification. In 1978, a coherent scheme was established in France, and the first certifications were issued under the responsibility of the COFREND (*). A French set of standards then described the terms and conditions for the certification of operators.

In 1994, the COFREND adopted the European standard EN 473 – Qualification and certification of END personnel, published by the European Committee for Standardization – *Comité Européen de Normalisation* (CEN). In 1996, after conforming to the European standard EN 45 013, the COFREND was accredited by the COFRAC (**) as a center of certification for individuals. Agreements between other national systems have since been established, notably at the European level. The set-up of the French and European certification systems was largely inspired by the American system ASNT (***) and its referential standard, the specification SNT TC 1A.

Since 2012, a standard of international scope EN ISO 9712 has replaced the EN 473, which defines the terms and conditions for the "qualification and certification of NDT personnel."

(*) COFREND: *Confédération française pour les essais Non Destructifs*, (French confederation for non-destructive testing)

(**) COFRAC: *Comité Français d'Accréditation* (French Committee for Accreditation)

(***) ASNT: *American Society for Non-Destructive Testing*

A.3   A FEW DEFINITIONS

Some terms need to be defined in order to ensure clear understanding of the logic of the certification scheme.

- **Qualification:** demonstration of physical attributes, knowledge, skill, training, and experience required to properly perform NDT tasks.
- **Certification:** procedure, used by the certification authority, to confirm that the qualification requirements for a method, level, and sector have been fulfilled, leading to the issuing of a certificate. It represents all of the measures and actions undertaken to ensure the qualification of a test operator in relation to a set of tasks and provides the operator with a valid certificate proving so.
- The certification authority offers the company, its clients, and contractors the means to guarantee and check the qualification of the test operators. This does not free the company (the employer) from its responsibility to authorize the test operator to execute the test and to guarantee its result. The certification validity is limited in time and functions under the condition that there is evidence (at the discretion of the employer) that the holder really and periodically practices and that he or she is still able to meet the requirements of the entire

exam, or a determined part, that he or she had to pass to be certified. Certification cannot be considered as a diploma in an academic sense.

- **Non-destructive testing method:** application of a physical principle in an NDT.
- **Non-destructive testing technique:** specific use of an NDT method.
- **Non-destructive testing instruction:** written description listing the precise steps to be followed when executing a test according to a recognized text: standard, code, specification, or procedure for a non-destructive test.
- **Non-destructive testing procedure:** written description of all the essential parameters and precautions to take when a non-destructive test is carried out on products, in accordance with a standard or standards, code(s), or specification(s).
- **Non-destructive testing inspection operator:** an individual whose core professional activity is dedicated to the application of non-destructive testing methods in the industry, in compliance with the codes and standards in effect.
- **Field in which the certification applies:** this concerns one individual, the test operator, for a given level of qualification in a particular method and industrial sector (COFREND) or in a defined type of product (ASNT).
- **Three different levels of qualification**: level 1, level 2, and level 3.

A.4   MODES OF CERTIFICATION

A.4.1   Methods and symbols

Today, the European system, according to the EN ISO 9712 standard, lists 10 methods of control, each symbolized by the 2-letter acronym of the English designation:

METHODS	SYMBOL	SIGNIFICATION
Ultrasounds	UT	Ultrasonic Testing
Radiography	RT	Radiographic Testing
Magnetoscopy	MT	Magnetic Testing
Dye Penetrant Testing	PT	Penetrant Testing
Eddy Current	ET	Eddy current Testing
Leakage	LT	Leak Testing
Visual Inspection	VT	Visual Testing
Acoustic Emission	AT	Acoustic Testing
Infrared Thermography	TT	Thermal Testing
Residual Strength Inspection	ST	Strength Testing

Note: Some additional methods are very specific to their applications or rely on a technology that is the exclusive property of a manufacturer and thus are the subject of independent certifications that are strongly inspired by standards of the EN ISO 9712 type. These may become a part of the EN ISO 9712 standard in the future if their use is more widely developed.

A.4.2 Levels of competences

The different systems identify 3 levels, each one corresponding to different qualification requirements.

Level 1

A level 1 operator is entitled to perform non-destructive testing for the method and sector (or product) in question, according to a written instruction and under the responsibility of a level 2 or 3 operator. He or she can:
- perform equipment calibration;
- execute the tests;
- observe and classify the results according to criteria; and
- record the results.

A level 1 operator is a technician who is not entitled to choose the method or the technique used and who does not evaluate or interpret the results.

Level 2

A level 2 operator is able to perform non-destructive testing according to an accepted procedure for the method and sector (or product) in question. This operator can:
- choose the testing technique and define the limits of application of the method;
- write instructions based on standards and specifications;
- calibrate and check the equipment;
- execute and supervise tests;
- evaluate and interpret the results in accordance with standards, codes, etc.;
- write instructions for a level 1 operator;
- execute, monitor, and guide tasks of an equivalent level or lower; and
- write the test reports.

A level 2 operator is a technician who is given a large degree of independence but who cannot choose the method and has to conform to criteria from the reference documents, the client/contracting party, or recommendations given by a level 3 operator.

Level 3

The competences of a level 3 test operator cover a wider field. For the method in question, this person is entitled to perform and run any operation, that is:
- draw up and validate instructions and procedures;
- interpret standards and codes;
- determine the methods, techniques, and the use of references necessary for a non-destructive test; and
- perform and supervise the tasks executed by level 1 and 2 operators and guide personnel at every level.

A level 3 individual has to be able to take responsibility for an installation or an exam center in non-destructive testing. The required knowledge and experience covers materials science, manufacturing, and other non-destructive testing methods, which makes this person competent to choose the methods, determine the techniques, and take part in the definition of the acceptance criteria, if necessary.

Training

A candidate applying for certification has to prove their level of training in the relevant method. Generally, there is a minimum training time required and this depends on the method in question and on the qualification level. Accessing level 2 and 3 certifications (in accordance with the EN ISO 9712 standard) is possible either after a level 1 certification, or level 2, or directly. To access the certification directly, the training time comprises the duration of the concerned level and the duration of the "lower" level.

For example, in ultrasonic testing, the duration of training is 40 hours for level 1, 80 hours for level 2, and 40 hours for level 3. In the case of direct access:

- At level 2, training duration must be 40 + 80 = 120 hours
- At level 3, training duration must be 40 + 80 + 40 = 160 hours

These durations can be reduced for candidates applying for several certifications and for graduates from technical higher education.

Certification bodies issue recommendations to the companies and training bodies about the programs and content related to the preparation of operators for the qualification exams.

Industrial experience

A candidate applying for certification has to prove a minimum of industrial experience, which varies according to the method and level of certification in question. As for training duration, the required experience for direct access to level 2 or level 3 is that required for the concerned level plus that of the "lower" level.

For example, in ultrasonic testing, the required experience is 3 months for level 1, 9 months for level 2, and 18 months for level 3. In the case of direct access:

- At level 2, experience must be: 3 + 9 = 12 months
- At level 3, experience must be: 9 + 18 = 27 months

Qualification exam

For levels 1 and 2, the exam is based on a general questionnaire on the method, a specific questionnaire on the application, and a practical exam. The level 3 exam is composed of a basic questionnaire, a questionnaire on the method in question, and a piece of work on documentation.

Moreover, and whatever the method and level, the candidate for certification has to prove his or her sight is satisfactory, allowing the:

- Ability to read at a minimum the Figure 1 on the Jaeger scale, or the letter N in "Times Roman" font, size 4.5, or in an equivalent font
- Proper color vision with the ability to distinguish contrasts, in colors or shades of gray, adapted to the method in question

The Table below recapitulates the exam terms and conditions. To pass the exam, the candidate must obtain at least a 70% grade in each part of the exam.

TERMS	LEVEL 1	LEVEL 2	LEVEL 3
Number of questions in the general exam	30 or 40 according to the method		Basic exam: 25 (materials, processes) + 10 (qualification/ certification)
Number of questions in the specific exam	20		+ 15 for the other 4 methods, i.e., 60 (level 2 questions)
Practical exam	Reading the instructions + testing of 2 test specimens (in general) + Report	Testing of 2 test specimens (in general) + interpretation + writing instructions	Exam in the main method: 30 (general) + 20 (specific to the relevant sector) Writing or analyzing a procedure

Renewal and certification validity

At the end of a first period of validity (10 years), certification can be renewed for a period of 5 years if the candidate can justify with proper documents that he or she has practiced, without interruption, the activity of non-destructive test operator for the method in question. A satisfactory level of vision needs to be proven again.

If the candidate for certification has interrupted their work practice for a significant period of time, he or she has to undergo a procedure of re-certification essentially based on a practical exam in accordance with a simplified procedure. To pass the exam the candidate must obtain at least a 70% grade.

A.5   A COMPARISON OF THE CERTIFICATION SYSTEMS

European certifications (like the French COFREND), are very similar in terms and conditions and with regard to the content of the exam to the Anglo-Saxon ASNT system. The difference comes essentially from the administrative organization of the certification scheme, from the prerogatives of the certification holder, and from the fields concerned by the certification.

The COFREND is a central certification authority, though there are sectorial committees. The certification is decided and issued by the COFREND. The ASNT is a company certification, which also delegates power to decide certification to institutions or organizations according to their own internal procedures. A COFREND certification belongs to a candidate who passed the exam, the holder keeps the prerogatives of the certification when changing or quitting an employer.

An ASNT certification belongs to the employer for which the candidate was working at the moment it was issued. The ASNT or the delegate authority issuing the certification can only transfer it to the new employer of the test operator if the original employer authorizes the transfer. The ASNT more commonly issues certifications of limited applications, on a given technique related to a method, or for a type of product on which the method is applicable. The COFREND certifications concern a method in its global applications.

INDEX